AS UNIT 1

STUDENT GUIDE

CCEA

Biology

Molecules and cells

John Campton

Philip Allan, an imprint of Hodder Education, an Hachette UK company, Blenheim Court, George Street, Banbury, Oxfordshire OX16 5BH

Orders

Bookpoint Ltd, 130 Park Drive, Milton Park, Abingdon, Oxfordshire OX14 4SE

tel: 01235 827827

fax: 01235 400401

e-mail: education@bookpoint.co.uk

Lines are open 9.00 a.m.–5.00 p.m., Monday to Saturday, with a 24-hour message answering service. You can also order through the Hodder Education website: www.hoddereducation.co.uk

ISBN 978-1-4718-6300-4

First printed 2016

Impression number 6

Year 2020

This guide has been written specifically to support students preparing for the CCEA AS and A-level Biology examinations. The content has been neither approved nor endorsed by CCEA and remains the sole responsibility of the author.

Cover photo: andamanse/Fotolia; p. 79 Dr Gopal Murti/Science Photo Library; p. 92 Biophoto Associates/Science Photo Library

Typeset by Integra Software Services Pvt. Ltd, Pondicherry, India

Printed in India

Hachette UK's policy is to use papers that are natural, renewable and recyclable products and made from wood grown in sustainable forests. The logging and manufacturing processes are expected to conform to the environmental regulations of the country of origin.

Contents

Content Guidance

Questions & Answers

Section A Structured questions

Section B Essay questions

Getting the most from this book

Exam tips

Advice on key points in the text to help you learn and recall content, avoid pitfalls, and polish your exam technique in order to boost your grade.

Knowledge check

Rapid-fire questions throughout the Content Guidance section to check your understanding.

Knowledge check answers

1 Turn to the back of the book for the Knowledge check answers.

Summaries

- Each core topic is rounded off by a bullet-list summary for quick-check reference of what you need to know.

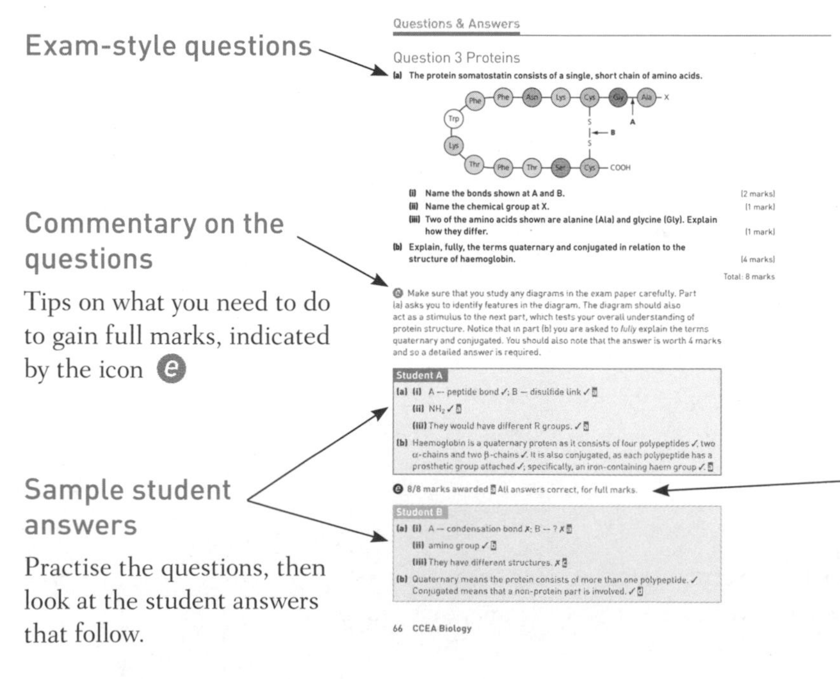

Commentary on sample student answers

Read the comments (preceded by the icon e) showing how many marks each answer would be awarded in the exam and where marks are gained or lost.

About this book

The aim of this book is to help you prepare for the AS Unit 1 examination for CCEA Biology. It also offers support to students studying A2 biology, since topics at A2 rely on an understanding of AS material.

The **Content Guidance** contains everything that you should learn to cover the specification content of AS Unit 1. It should be used as a study aid as you meet each topic, for end-of-topic tests, and during your final revision. For each topic there are *exam tips* and *knowledge checks* in the margins. Answers to the knowledge checks are provided towards the end of the book. At the end of each topic there is a list of the practical work with which you are expected to be familiar. This is followed by a comprehensive, yet succinct, summary of the points covered in each topic.

The **Questions & Answers** section contains questions on each topic. There are answers written by two students with comments on their performances and how they might have been improved. These will be particularly useful during your final revision. There is a range of question styles to reflect those you will encounter in the AS Unit 1 exam, and the students' answers and comments will help with your examination technique.

Developing your understanding

It is important that through your AS course you develop effective study techniques.

- You must *not* simply read through the content of this book.
- Your understanding will be better if you are *active* in your learning. For example, you can take the information given in this book and present it in different ways:
  - a series of bullet points to summarise the key points
  - annotated diagrams to show structure and function (e.g. a diagram of a cell with labelled features and brief notes on function) or annotated graphs (e.g. showing enzyme properties with notes attached explaining the trends)
  - a spider diagram, e.g. one on enzymes would include reference to theoretical aspects, the effects of temperature and pH, the effect of inhibitors and the effect of immobilisation
- Compile a glossary of terms for each topic. Key terms in this guide are shown in **bold** (with some defined in the margin) and for each you should be able to provide a definition. This will develop your understanding of the language used in biology and help you where *quality of written communication* is being assessed.
- Write essays on different topics. For example, an essay on enzymes will test your understanding of the entire topic and give you practice for the Section B question.
- *Think* about the information in this book so you are able to *apply* your understanding in unfamiliar situations. Ultimately you will need to be able to deal with questions that set the topic in a new context.
- Ensure that you are familiar with the expected practical skills, as questions on these may be included in this unit.
- Use past questions and other exercises to develop all the skills that examiners must test.
- Use the topic summaries to check that you have covered all the material that you need to know and as a brief survey of each topic.

Content Guidance

Biological molecules

The chemical composition of living organisms

Only 20 different types of atom (of the 92 stable elements) occur in living organisms. The elements present in the largest proportions are carbon (C), hydrogen (H), nitrogen (N), oxygen (O), phosphorus (P) and sulfur (S). These atoms have low atomic mass and combine with one another to form molecules (many of which may be compounds) held together by strong **covalent bonds**. Consequently, living organisms are both light and strong. The remaining atoms occur as charged ions (e.g. Ca^{2+}, Fe^{2+}, K^+ and NO_3^-).

Element A substance containing only one type of atom.

Molecule Two or more atoms chemically bonded.

Compound A molecule containing two or more different elements.

Ion An atom or group of atoms with an electric charge due to the loss or gain of one or more electrons.

Macromolecule A big molecule, such as a large carbohydrate (e.g. starch), lipid, protein or nucleic acid, consisting of smaller molecules linked together.

Solvent The liquid in which a solute is dissolved to form a solution. Ions and charged molecules (solutes) are surrounded by water (solvent) and so are separated into solution.

The **water content** of living organisms ranges from 50% to 95%. Water is composed of hydrogen and oxygen; its formula is H_2O. Water is a liquid whereas other substances of similar molecular mass are gases (e.g. CH_4, NH_3, O_2 and CO_2). The reason for this is that water is a **dipolar** molecule (the oxygen is slightly negative, δ–, and the hydrogen is slightly positive, δ+), so neighbouring water molecules are linked by **hydrogen bonds** — weak bonds between the oxygen on one water molecule and a hydrogen on another (Figure 1).

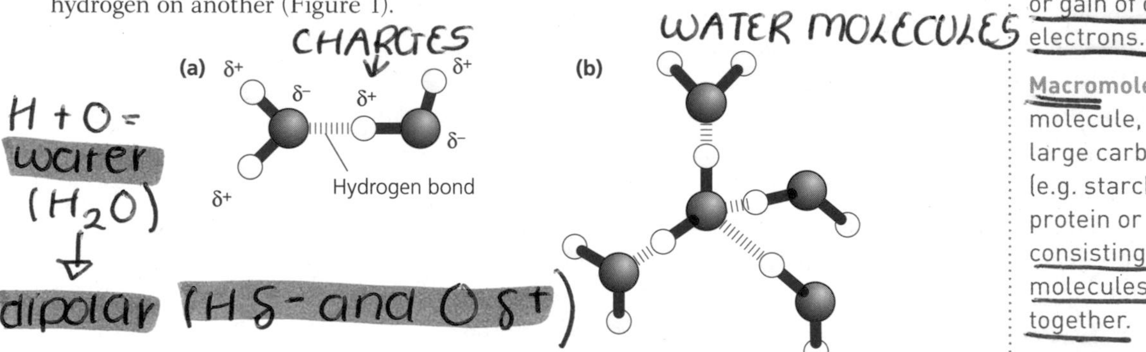

Figure 1 (a) The charges on water molecules. (b) A cluster of water molecules

The dry mass of living organisms is in the form of carbohydrates, lipids, proteins and nucleic acids, most of which are macromolecules.

Water is a good **solvent** capable of dissolving a wide range of chemical substances. This includes all ions and molecules with charged groups. Ions are charged and are surrounded by shells of water; water clusters around the charged groups of glucose and amino acids. This is shown in Figure 2.

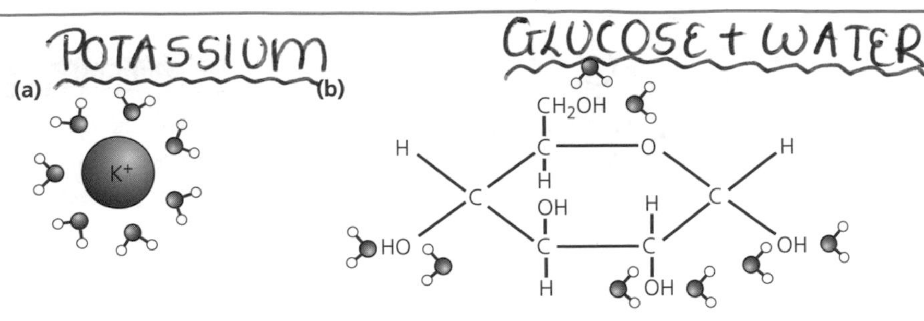

Figure 2 (a) K^+ with a shell of water molecules
(b) Glucose with clusters of water molecules

Knowledge check 1

a Explain how water acts as a solvent.

b What type of molecules will not dissolve in water?

Cohesion: H bonding of water molecules

Knowledge check 2

What type of energy can break hydrogen bonds easily?

The biochemical reactions within the cell are carried out in solution. (Since the water content of seeds and spores is as low as 10% their biochemical activity is suspended until they become rehydrated.) Water is also used to transport nutrients and waste substances.

The hydrogen bonding of water molecules is known as **cohesion**. This is an important property in the flow of a continuous column of water (and dissolved nutrients) through the xylem vessels of plants. You will study this in AS Unit 2.

Water also has an important role in temperature regulation since evaporation of water from a surface cools it down. The energy required to break the hydrogen bonding in liquid water is known as the **latent heat of evaporation**.

Since **ions** are soluble in water this is the way in which living organisms absorb certain elements. The atoms contained within the ion have specific roles (Table 1). Some ions maintain osmotic balance (e.g. potassium ions, K^+) while some act as pH buffers (e.g. hydrogen carbonate ions, HCO_3^-).

Table 1 The role of ions and their atoms as components of biologically important molecules

Ion	Chemical symbol	Role in biological molecules
Nitrate	NO_3^-	Nitrogen in the amino group of amino acids and in the organic bases of nucleotides produced in plants
Sulfate	SO_4^{2-}	Sulfur in the R group of the amino acid cysteine
Phosphate	PO_4^{3-}	In a range of important molecules: adenosine triphosphate, ATP; nucleotides and so nucleic acids; phospholipids
Calcium	Ca^{2+}	In calcium pectate, which contributes to the middle lamella of plant cell walls; in calcium phosphate in the bone of vertebrate animals
Magnesium	Mg^{2+}	In the chlorophyll molecule; in magnesium pectate; also present in the middle lamella of plant cell walls
Iron	Fe^{3+}	In the haemoglobin molecule

Since ions and polar molecules (e.g. glucose and amino acids) are charged and have shells or clusters of water, this influences how they pass through cell-surface membranes (pp. 42–45).

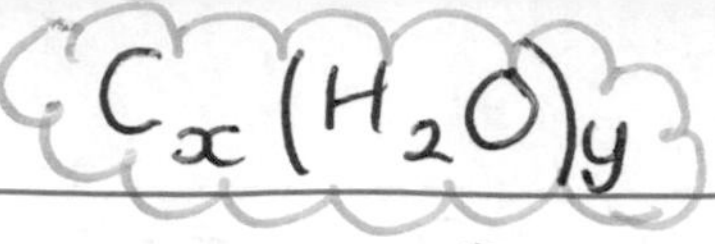

Carbohydrates C H O

Carbohydrates contain carbon, hydrogen and oxygen. They have the general formula $C_x(H_2O)_y$. The simplest carbohydrates are single sugars (**monosaccharides**) with the formula $(CH_2O)_n$, where n can vary from 3 to 9. The important types of monosaccharide are **trioses** ($n = 3$), **pentoses** ($n = 5$) and **hexoses** ($n = 6$). Two hexose sugars bond together to form a **disaccharide** (double sugar) via a **condensation** reaction (a chemical reaction in which two molecules are joined together and one molecule of water is released). **Polysaccharides** are formed when many hexose sugars are linked by condensation reactions. Disaccharides and polysaccharides release hexose sugars via **hydrolysis**.

Knowledge check 3

A heptose is a monosaccharide with seven carbon atoms. One example is sedoheptulose, which is produced during photosynthesis. What is its molecular formula?

Important monosaccharides include:

- the pentose sugars **ribose** and **deoxyribose** — these are constituents of nucleotides that form the nucleic acids RNA and DNA
- the hexose sugars α-glucose, β-glucose and fructose (Figure 3)

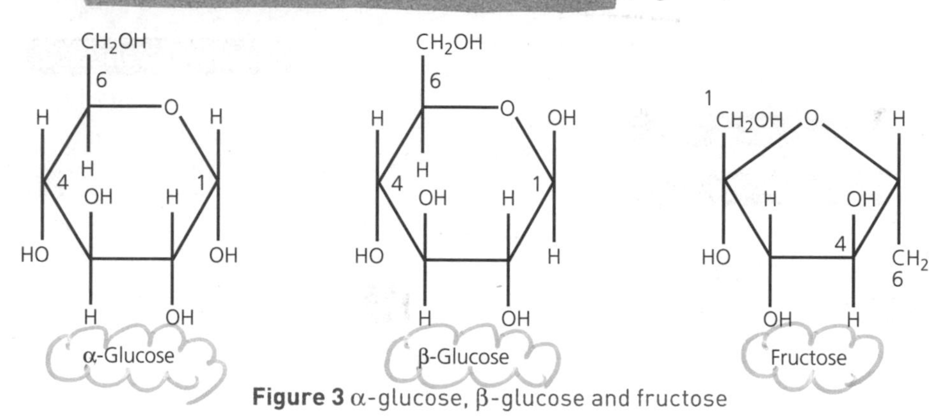

Figure 3 α-glucose, β-glucose and fructose

The carbon 1, carbon 4 and carbon 6 positions are indicated in the glucose molecules shown in Figure 3. These are important since it is at these positions that different glucose molecules bond together. The subtle but important distinction between α-glucose and β-glucose is that the –H and –OH groups at the carbon 1 position are reversed. This means that the two types of glucose bond slightly differently.

The hexose sugars can bond together to produce disaccharides:

- α-glucose and fructose form **sucrose** — sucrose (cane sugar) is the form in which carbohydrates are transported in plants
- α-glucose and α-glucose form **maltose** — the product of starch digestion

The formation of maltose is a **condensation reaction** since water is removed in the process. The bond formed is an **α-1,4-glycosidic bond** (Figure 4). Note that the two α-glucose molecules are not in the same plane.

The breaking of a glycosidic bond is a **hydrolysis reaction**. In this case, maltose would be hydrolysed to its constituent α-glucose molecules. The formation and hydrolysis of maltose are shown in Figure 4.

a-glucose + fructose → sucrose
a-glucose + a-glucose → maltose

Exam tip

Be careful not to confuse the terms 'maltose' and 'maltase'. The ending '-ose' tells you that the molecule is a carbohydrate; the ending '-ase' tells you that the molecule is an enzyme. Maltase is the enzyme that catalyses the hydrolysis of maltose.

α-Glucose

CH_2OH H H 4 1 HO OH

CH_2OH H H 4 1 HO OH

α-Glucose

Condensation H_2O

Hydrolysis H_2O

Maltose

CH_2OH H H 4 1 HO O CH_2OH H H 4 1 OH

α-1,4-glycosidic bond

Note that the bond causes the **α**-glucoses to lie at different angles

Figure 4 The formation and hydrolysis of maltose

Many α-glucose molecules can bond to produce polysaccharides, which are examples of **polymers**. The simplest is **amylose**, a constituent of starch in plants (Figure 5). Only α-1,4-glycosidic bonds are involved and a helical structure results since the α-glucoses join at slightly different angles to each other.

Polymer A chain-like macromolecule made up of many similar units (called monomers) bonded together. All polysaccharides are polymers of monosaccharides (e.g. glucose).

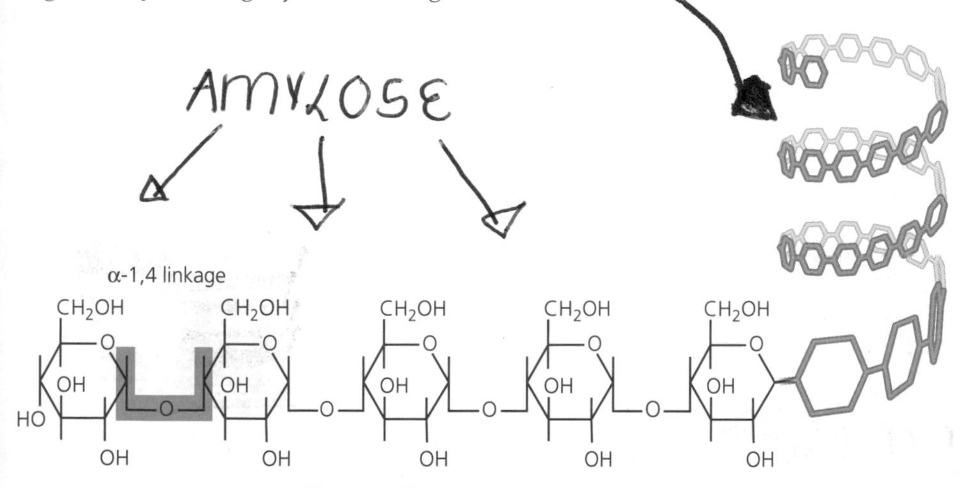

Figure 5 The structure of amylose

Two other polysaccharides are produced when **α-1,6-glycosidic bonds** are added at regular intervals to produce branches. In **amylopectin**, also a constituent of starch, α-1,6 bonds occur every 24–30 glucose units; in **glycogen**, found in the liver and muscle cells of mammals, α-1,6 bonds occur every 8–10 glucose units, resulting in more frequent branching. The structure of a branched polysaccharide is shown in Figure 6.

Knowledge check 4

Identify two ways in which starch molecules from different plants can differ from each other.

BRANCHED MOLECULES

Amylose & amylopectin = energy stores

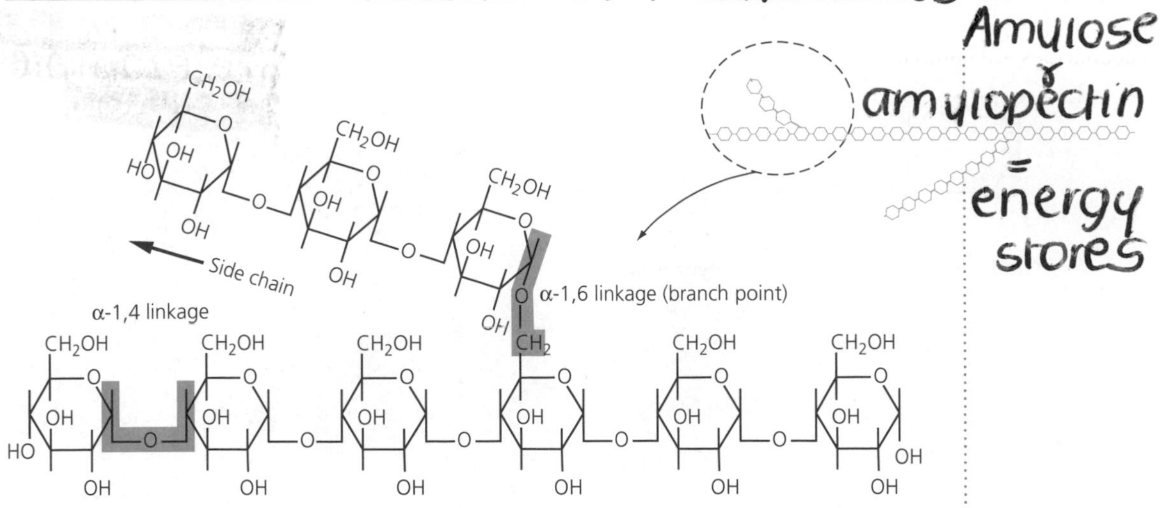

Figure 6 The structure of branched polysaccharides: amylopectin and glycogen

Glucose is an important respiratory substrate. The polysaccharides amylose and amylopectin in starch, and glycogen, are the means by which glucose is stored in organisms (so they are regarded as energy stores). They are well adapted as storage molecules because:

- many glucose molecules can be stored in a cell (amylose molecules are coiled and so are compact)
- they are readily hydrolysed to release glucose molecules (especially amylopectin and glycogen, which are branched and so have many terminal ends on which hydrolytic enzymes may act)
- they are insoluble and so cannot move out of cells
- they are osmotically inert (only soluble substances affect the water potential of a cell — pp. 44–45)

Knowledge check 5

Glycogen is significantly more branched than amylopectin. Explain how this difference is important in animals, which have a much greater metabolic rate than plants.

Many β-glucose molecules bond to produce the polysaccharide **cellulose**. Due to the way in which the β-glucose molecules are orientated cellulose molecules are straight chains. Adjacent cellulose molecules are linked by hydrogen bonds and grouped to produce **microfibrils**. Each microfibril consists of hundreds of cellulose molecules and is a structure of immense **tensile strength**. Cellulose microfibrils mesh to form the **cell wall** of the plant cell and prevent the cell from bursting in dilute solutions. The structure of cellulose is shown in Figure 7.

Tensile strength The ability of a material to withstand being stretched (pulled) before breaking.

Cellulose molecules linked by H bonds to produce microfibrils.

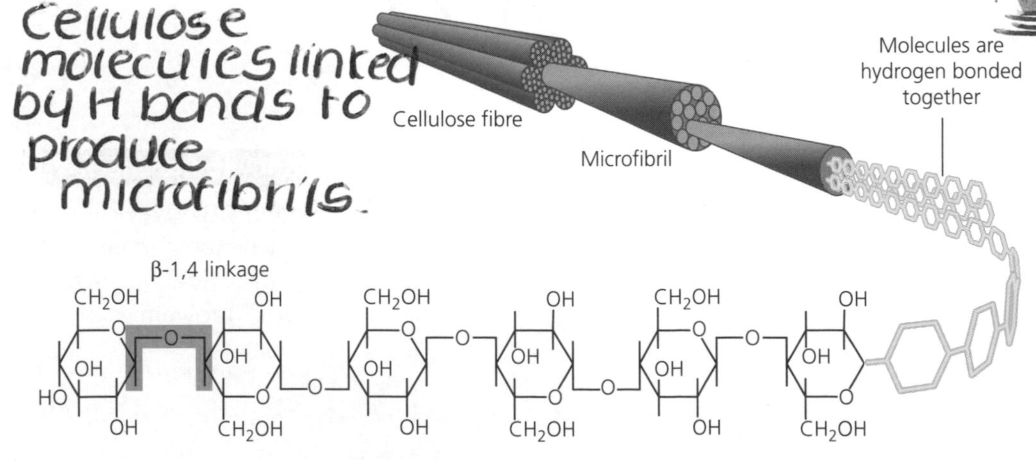

Figure 7 The structure of cellulose

Some polysaccharides are compared in Table 2.

Table 2 A summary of different polysaccharides

Polysaccharide	Monomer	Glycosidic bond(s)	Shape of polymer	Location	Function
Amylose	α-glucose	α-1,4 bonds only	Unbranched, helical molecule	Starch grains in living plant cells	Store of glucose (energy)
Amylopectin	α-glucose	α-1,4 and α-1,6 bonds	Branched, helical molecule	Starch grains in living plant cells	Store of glucose (energy)
Glycogen	α-glucose	α-1,4 and α-1,6 bonds	Branched, helical molecule	Granules in liver and muscle cells of mammals	Store of glucose (energy)
Cellulose	β-glucose	β-1,4 bonds only	Straight chains cross-linked to parallel chains	Cell walls of plant cells	Structural support to plant cell

Knowledge check 6

Give two structural differences between amylose and cellulose.

Lipids

Lipids contain mostly carbon and hydrogen with a few atoms of oxygen; they may contain other types of atom. They are macromolecules, but they are not polymers. They are a diverse group structurally, the common feature being their insolubility in water. They can be extracted from cells by organic solvents. Examples include triglycerides (fats and oils), phospholipids, steroids (e.g. cholesterol) and waxes.

Exam tip

Be careful not to confuse the terms 'fatty acid' and 'fat'. A fatty acid is one product of lipid hydrolysis; a fat is a lipid in solid form.

Fatty acids are an important constituent of triglycerides and phospholipids. They are essentially long hydrocarbon chains with a carboxylic acid group at one end. They can vary according to:

- the length of the hydrocarbon chain
- whether the hydrocarbon chain contains double bonds — a hydrocarbon with double bonds is described as **unsaturated** in comparison with chains with only single bonds that are **saturated** with hydrogen (Figure 8)

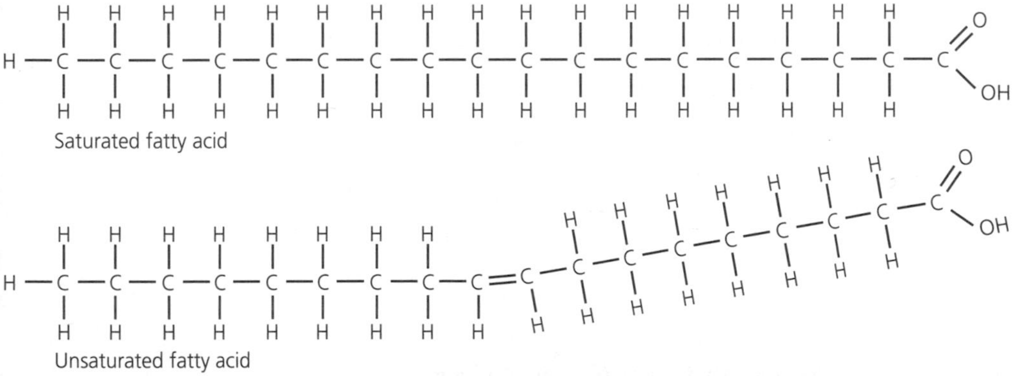

Saturated fatty acid

Unsaturated fatty acid

Figure 8 A saturated fatty acid and an unsaturated fatty acid with one double bond

Triglycerides consist of a **glycerol** molecule bonded by condensation reactions to three fatty acids. **Ester bonds** are formed. The constituent molecules of a triglyceride are released by hydrolysis reactions. The formation and hydrolysis of a triglyceride are shown in Figure 9.

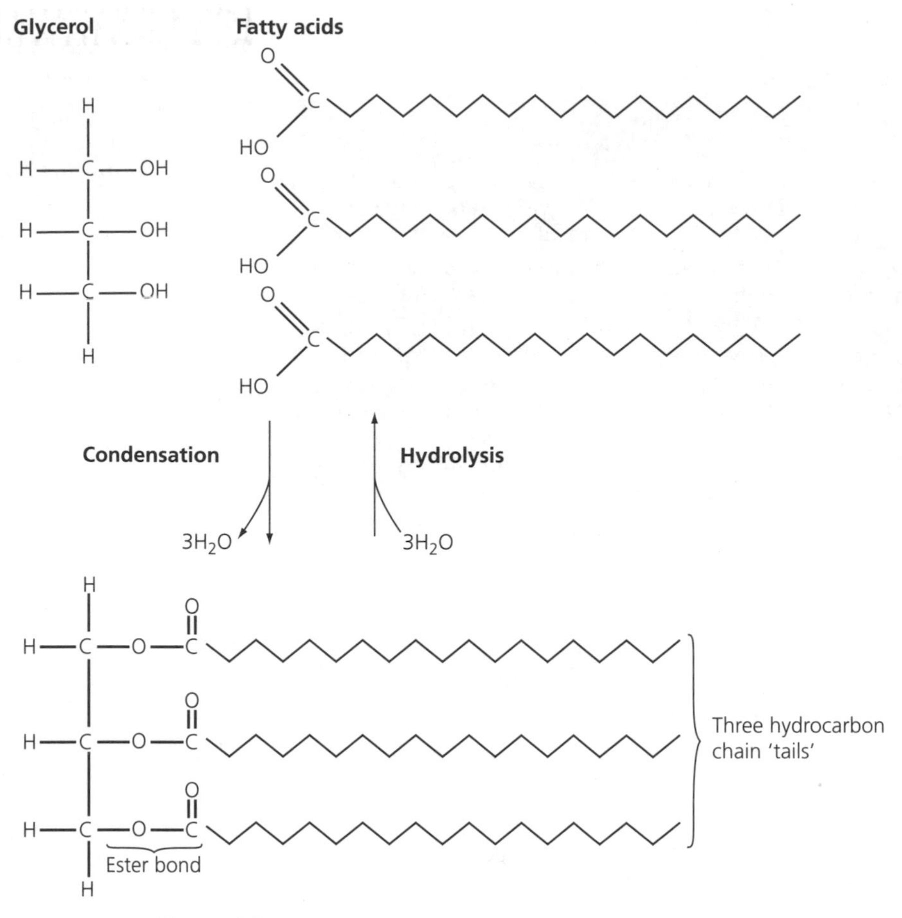

Figure 9 The formation and hydrolysis of a triglyceride

Triglycerides with unsaturated hydrocarbon chains (and/or with shorter chains) have lower boiling points than those with saturated hydrocarbon chains. Therefore:

- triglycerides with unsaturated hydrocarbon chains (and shorter chains) tend to be liquid at room temperatures — oils
- triglycerides with saturated hydrocarbon chains (and longer chains) are solid — fats

Oils tend to be found in plants while fats occur in animals.

Like polysaccharides, triglycerides represent energy stores (particularly the constituent fatty acids). They are energy-rich and, gram-for-gram, they release more energy than carbohydrates. As a mass-efficient means of storing energy, lipids are found in seeds (e.g. linseed oil), migratory birds (e.g. geese) and in the camel's hump.

Fatty acids may only be respired aerobically.

Fats are also important in providing:

- a thermal insulating layer in mammals, since they are poor heat conductors
- buoyancy in marine mammals such as dolphins and whales
- a cushioning layer around and, therefore, protection to internal organs such as the kidneys
- water when respired — this is metabolic water, which is important for desert animals such as gerbils and camels

Knowledge check 7

Triglycerides are macromolecules but are not polymers. Explain why triglycerides are not polymers.

Exam tip

The initial letters SSS and DUL are useful memory aids: SSS — **S**ingle (bonds), **S**aturated (triglyceride), **S**olid (at room temperature); DUL — **D**ouble (bonds), **U**nsaturated (triglyceride), **L**iquid (at room temperature).

A **phospholipid** consists of a glycerol molecule, two fatty acid residues and a phosphate group (Figure 10). The phosphate causes the glycerol end (the 'head') to be polarised and, therefore, soluble in water (**hydrophilic** or 'water-loving'); the long hydrocarbon chains (the 'tails') are non-polar and insoluble in water (**hydrophobic** or 'water-hating').

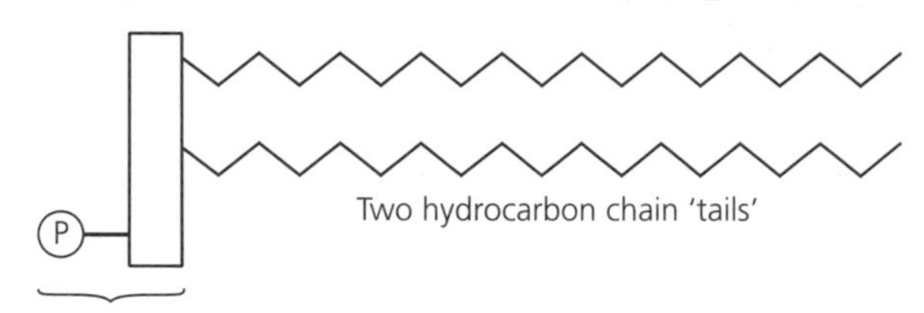

Figure 10 A phospholipid

Knowledge check 8

Distinguish between a triglyceride and a phospholipid.

In an aqueous environment, phospholipids automatically form bilayers. The phospholipid bilayer represents the basis of membrane structure in the cell (pp. 40–41).

The steroid **cholesterol** is essentially a molecule with a hydrocarbon chain and four carbon-based rings. It is found in cell membranes and, since it is hydrophobic, is located among the hydrocarbon chains of the phospholipid bilayer. A number of steroid hormones are synthesised from cholesterol, including the sex hormones oestrogen and testosterone.

Proteins

Proteins make up about two-thirds of the total dry mass of a cell. They contain carbon, hydrogen, oxygen, nitrogen and usually sulfur. Proteins are chains of amino acids. Since there are 20 different amino acids, which can be arranged in many different sequences, a huge variety of proteins is possible. Proteins have a highly organised structure with up to four levels of organisation. The overall shape is precise and integral to the function of the protein in the cell.

Amino acids consist of a carbon atom with four groups attached:

- an amino group
- a carboxylic acid group
- a hydrogen atom
- a residue (R-group)

It is the residue that differs to form the 20 different naturally occurring amino acids. Some of the residues carry a charge and so may be involved in hydrogen bonding, some are hydrophobic and a few contain sulfur (e.g. cysteine). The general structure of an amino acid is shown in Figure 11.

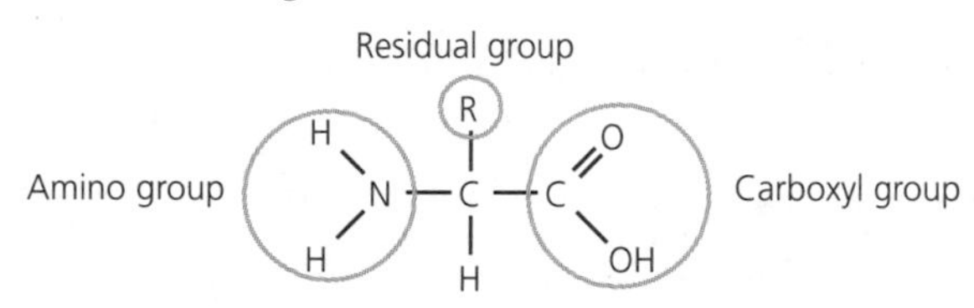

Figure 11 The general structure of an amino acid

Exam tip

With the exception of a generalised amino acid, you will not be asked to draw molecular structures from scratch. However, you may be required to recognise structures, or complete diagrams of carbohydrates, lipids or protein molecules, or their constituents.

Amino acids can bond together to form a **dipeptide**. A condensation reaction is involved and the amino acids are linked by a **covalent** **peptide bond**. A hydrolysis reaction breaks the dipeptide down to release the two amino acids. The formation and hydrolysis of a dipeptide are shown in Figure 12.

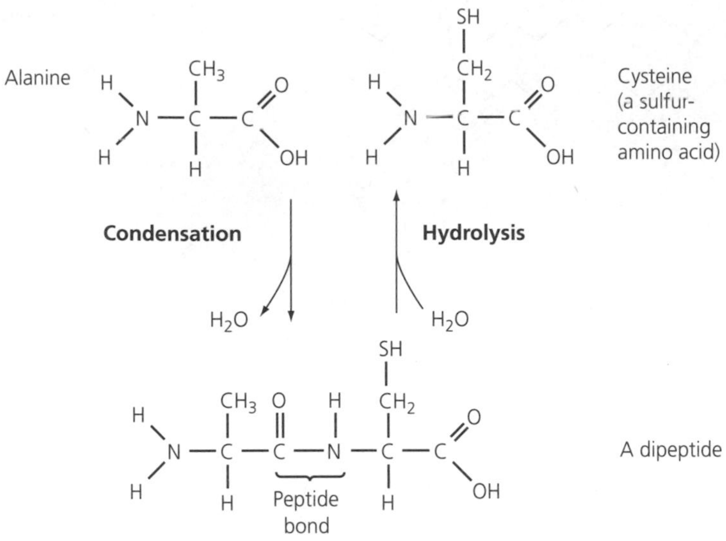

Alanine

Cysteine (a sulfur-containing amino acid)

Condensation

Hydrolysis

H_2O

H_2O

A dipeptide

Peptide bond

Figure 12 The formation and hydrolysis of a dipeptide

Many amino acids are peptide-bonded together to form a **polypeptide**. The **primary structure** of a polypeptide is the sequence of amino acids in the chain. The polypeptide has an amino group at one end and a carboxyl group at the other.

The **secondary structure** is either an α-helix or a β-pleated sheet. The structures are held in place by hydrogen bonds between peptide links in adjacent parts of the chain.

Globular proteins have a **tertiary structure**. The polypeptide folds over on itself in a precise way to produce a specific three-dimensional shape. This is due to interaction between the free R-groups of the amino acids. Different R-groups produce specific links with each other: hydrogen bonds between polar R groups; hydrophobic interactions between non-polar R-groups; ionic bonds between ionised R-groups; and disulfide bonds between the sulfur-containing R-groups of cysteine residues. A different order of amino acids means that the R-group interactions are different and so a different three-dimensional shape is generated. Fibrous proteins lack a tertiary structure.

Some proteins consist of two or more polypeptide chains bonded together. This is the **quaternary structure**.

The levels of organisation in a protein molecule are shown in Figure 13.

Covalent bond Bond formed when two atoms share one or more electrons. In proteins, peptide bonds and disulfide bonds are examples of covalent bonds. Covalent bonds are stronger than hydrogen bonds.

Exam tip

Be careful to distinguish the terms polypeptide and protein. A polypeptide represents a protein if it lacks a quaternary structure; however, if the finalised protein has a quaternary structure then it consists of more than one polypeptide.

Exam tip

The tertiary structure is formed as a result of interactions involving the R-groups of amino acids within a single polypeptide chain. The quaternary structure is the result of interactions between different polypeptides and occurs in the Golgi body as polypeptides arrive there in vesicles from the rough ER, where polypeptides are synthesised.

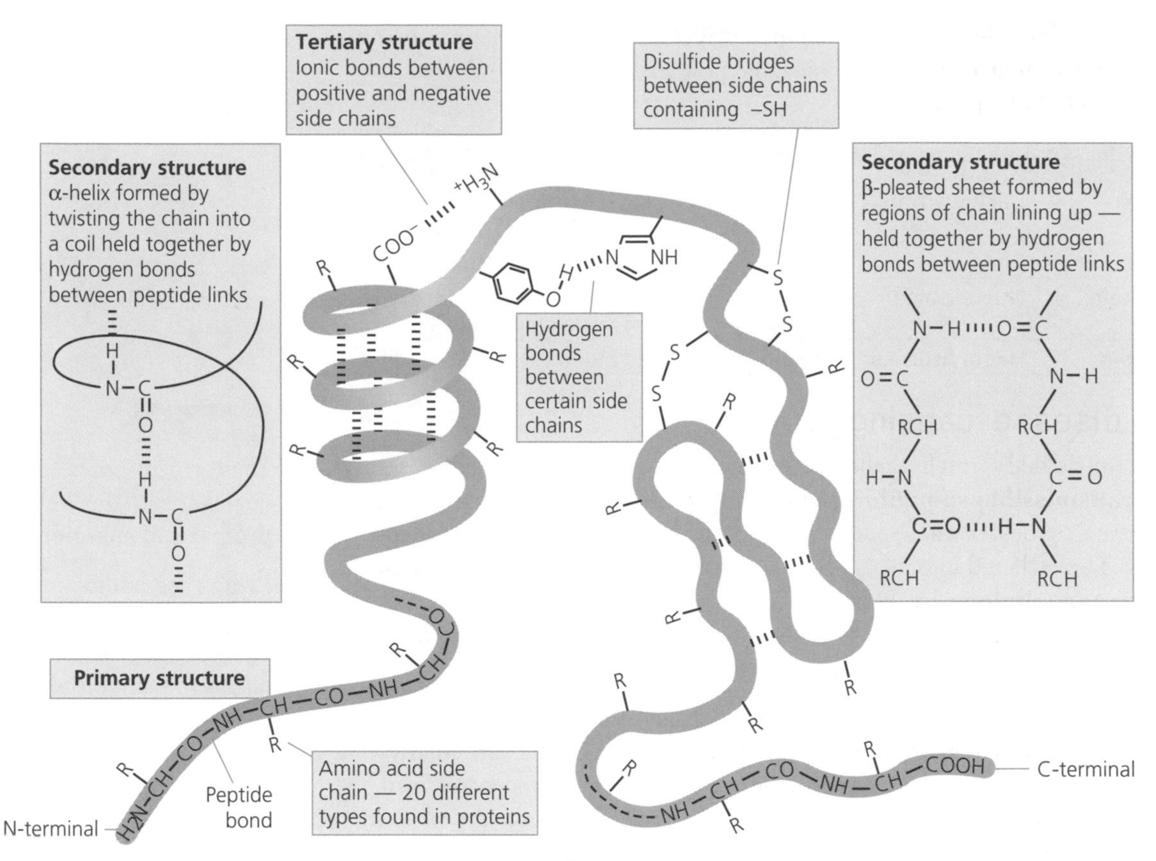

Figure 13 The different levels of structure in a protein molecule

Haemoglobin is a **globular** protein found in large quantities in red blood cells. Each molecule consists of four polypeptides: two α-chains and two β-chains (Figure 14a). Each polypeptide has an iron-containing haem group attached. Haemoglobin is important in the transport of oxygen in animals. An oxygen molecule can associate with each haem to form oxyhaemoglobin.

Collagen is a **fibrous** protein. Each molecule consists of three similar polypeptides coiled round each other and held together by hydrogen bonds (Figure 14b). Collagen molecules are bonded together to form the strong fibres found in the skin, tendons and ligaments.

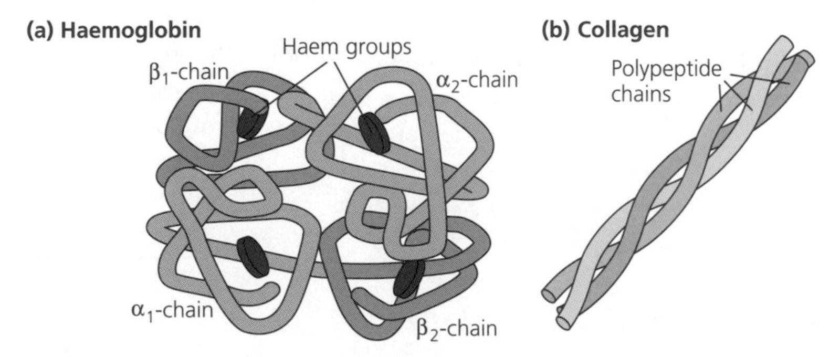

Figure 14 Haemoglobin (a) and collagen (b) are quaternary proteins

Knowledge check 9

List the types of bond that form each of the following levels of protein structure: primary structure, secondary structure and tertiary structure.

Knowledge check 10

Describe the similarities and differences between haemoglobin and collagen.

Conjugated proteins have a non-protein part attached. The non-protein component is called a **prosthetic group**. Some conjugated proteins are given in Table 3.

Table 3 Some important conjugated proteins

Name	Prosthetic group	Location
Glycoprotein	Carbohydrate	Mucin (component of saliva); cell-surface membrane
Lipoprotein	Lipid	Membrane structure
Nucleoprotein	Nucleic acid	Chromosome structure; ribosome structure
Haemoglobin	Haem (iron-containing)	Red blood cells

Prions: disease-causing proteins

Malformed (misfolded) proteins called **prions** are the agents that cause diseases known as **transmissible spongiform encephalopathies** (TSE) — diseases in which the brain tissue becomes spongy with holes where there were once groups of neurones. The afflicted mammal — sheep (with **scrapie**), cattle (with **bovine spongiform encephalopathy**, BSE, or 'mad-cow disease') or human (with **Creutzfeldt-Jacob disease**, CJD) — loses physical coordination. In humans, memory and body control are lost prior to death.

These diseases involve the replacement of a normal, cell-surface glycoprotein (PrP^C — C for cellular) that is located on the cell surface of a variety of cells, particularly neurones within the brain and spinal cord, with a structurally altered prion form (PrP^{Sc} — Sc for scrapie, the earliest known prion disease). The secondary structure of PrP^C is composed mainly of α-helices, whereas PrP^{Sc} is rich in β-sheets (Figure 15).

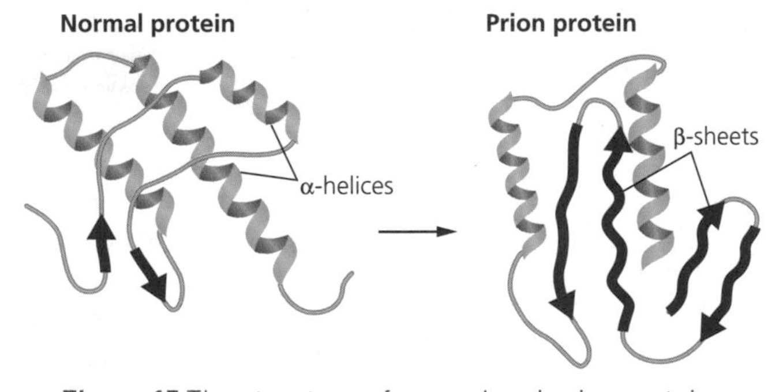

Figure 15 The structure of normal and prion proteins

The presence of disease protein (PrP^{Sc}) in the body causes the normal protein (PrP^C) to alter its secondary structure. This transformation of the normal protein sets up a chain reaction, which leads to a progressive accumulation of the prion protein on neurone surfaces, causing neurone degeneration and cell death. Prions are stable, being resistant to extreme temperatures and radiation. Prion disease is rare, but currently fatal.

Exam tip

The term 'prion' is derived from the words **pro**tein and infect**ion**.

Prions are pathogens (disease-causing agents) and are different from other pathogens, such as viruses and bacteria, which possess nucleic acid. It is the nucleic acid in viruses and bacteria that allows them to reproduce. Prions replicate by converting the normal PrP^C into the abnormal PrP^{Sc}.

Knowledge check 11

Explain why heat sterilisation of surgical instruments would not prevent transmission of CJD if they had previously been used during an operation on an infected patient.

Prion diseases arise in three different ways:

- **Transmission** through consumption of infected food, via an open wound exposed to the environment (prions can remain in the soil for many years) or as a result of a medical procedure involving infected material (such as tissue transplantation from an infected person). There is evidence that sheep prions may infect cattle and that cattle prions can infect humans (through eating infected meat from cattle).
- **Inheritance** of a gene mutation that would have occurred during meiosis in the production of an egg or sperm cell. The mutant gene codes for the synthesis of the misshapen prion protein, PrP^{Sc}. Years later (for CJD in humans up to 30 years) prion disease results when sufficient numbers of neurones have been damaged. Inherited prion disease accounts for about 15% of CJD cases.
- **Sporadically**, whereby the normal protein PrP^{C} spontaneously transforms into the disease form, PrP^{Sc}. The disease is regarded as being sporadic if there is no evidence of it being inherited or transmitted via an external source. Where it occurs in humans, the average age of onset is around 65. Sporadic prion disease accounts for 80% of all cases of CJD.

Nucleic acids

There are two types of nucleic acid: **deoxyribonucleic acid** (**DNA**) and **ribonucleic acid** (**RNA**). DNA carries the **genetic code** (for the synthesis of polypeptides) and is capable of **self-replication** so allowing genetic information to be passed from generation to generation. RNA molecules assist the functioning of DNA.

Nucleotide structure

Nucleotides are the subunits of nucleic acids. Each nucleotide (Figure 16) consists of:

- a **pentose** sugar — **deoxyribose** or **ribose**
- a **nitrogenous base** — from **adenine**, **guanine**, **cytosine** and **thymine** (in deoxyribonucleotides only) or **uracil** (in ribonucleotides only)
- a **phosphate** group (attached to the carbon-5 of the sugar)

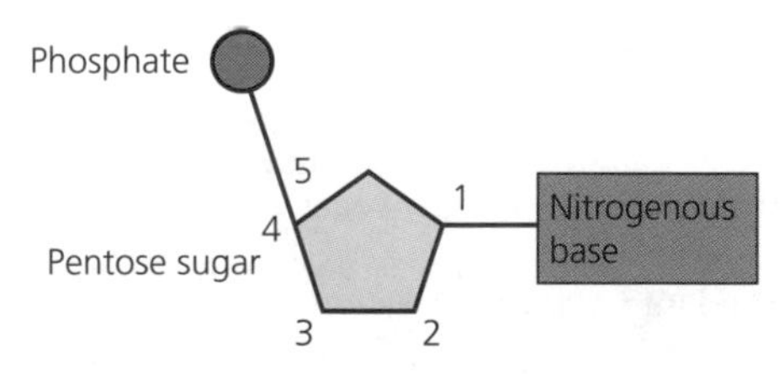

Figure 16 A generalised nucleotide (the numbers indicate the position of carbon atoms in the 5-carbon sugar)

Knowledge check 12

The components of a nucleotide are bonded by condensation reactions. How many molecules of water are produced during the formation of a nucleotide?

Deoxyribonucleotides and ribonucleotides are compared in Table 4.

Table 4 A comparison of deoxyribonucleotides and ribonucleotides

Feature	Deoxyribonucleotides	Ribonucleotides
Pentose sugar	Deoxyribose	Ribose
Nitrogenous base	Adenine (A), guanine (G), cytosine (C), thymine (T)	Adenine (A), guanine (G), cytosine (C), uracil (U)
Macromolecule formed	DNA	RNA

Nucleic acid structure

Nucleotides join together by condensation reactions forming **phosphodiester bonds** (between the phosphate of one nucleotide and the C3 of the pentose of the other nucleotide) along a 'sugar–phosphate' backbone. The polynucleotide strand formed has a free **5′-end** (with phosphate attached) and a free **3′-end** (Figure 17).

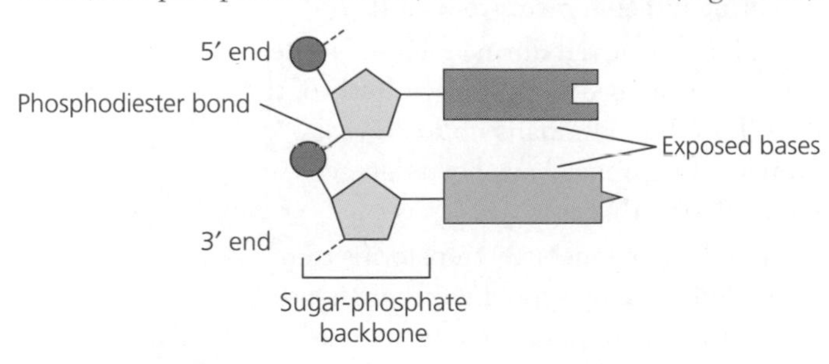

Figure 17 The basic structure of a polynucleotide

A DNA molecule consists of two **anti-parallel** strands. The bases on opposite strands are joined by hydrogen bonds in a precise way: adenine always bonds with thymine; guanine always bonds with cytosine (Figure 18). This is known as base pairing.

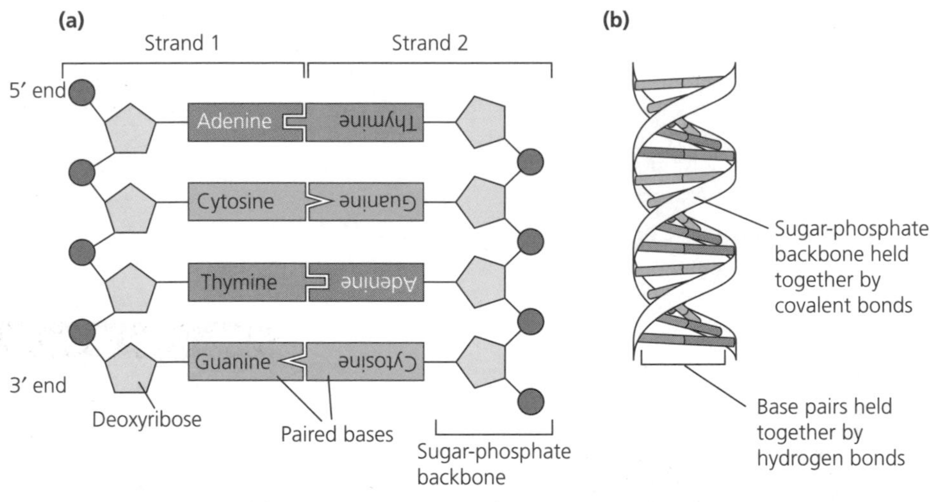

Figure 18 (a) DNA structure and (b) formed into a double helix

The nucleotides join at slightly different angles to each other and so the whole structure forms a **double helix**; the hydrogen bonding between the two strands increases its stability.

RNA molecules are single stranded and are much shorter than DNA molecules — DNA may be millions of nucleotides long; RNA usually consists of a few hundred nucleotides. There are three forms of RNA and one form, ribosomal RNA (rRNA), is a component of ribosomes.

DNA and RNA are compared in Table 5.

Knowledge check 13

In a nucleic acid polymer, which component parts are at the ends of the molecule?

Knowledge check 14

Define the term 'anti-parallel'.

Exam tip

If you are given the percentage of one base in a DNA molecule, you can readily calculate the percentages of the remaining bases. For example, if the bases of a DNA are 30% thymine, and adenine pairs with T, then there must be 30% A; and since the remaining 40% must be shared equally by cytosine and guanine (since they base pair) then there is 20% C and 20% G.

Exam tip

Be careful to distinguish the terms nucleotide and nucleic acid. For example, you could be asked to list differences between deoxyribonucleotides and ribonucleotides or between deoxyribonucleic acid and ribonucleic acid. There are more differences between DNA and RNA than between their constituent nucleotides — compare Tables 4 and 5.

Table 5 A comparison of DNA and RNA

Feature	DNA	RNA
Subunits	Deoxyribonucleotides (contains deoxyribose and thymine)	Ribonucleotides (contains ribose and uracil)
Length	Very long	Relatively short
Strands	Double stranded	Single stranded
Base pairing	A with T and G with C	No base pairing

DNA replication

Since DNA is the genetic code, it must be copied exactly from one generation to the next. This is achieved by self-replication, using a **semi-conservative mechanism** in which each strand acts as a template for the synthesis of a new strand. Each new DNA molecule contains one of the original strands in addition to a new strand (hence the name semi-conservative).

The sequence of events in DNA replication is as follows:

1 The enzyme DNA helicase breaks the hydrogen bonds holding the base pairs together and 'unzips' part of the DNA double helix, revealing two strands.

2 The enzyme DNA polymerase moves along each strand, which acts as a template for the synthesis of a new strand.

3 DNA polymerase catalyses the joining of free deoxyribonucleotides to each of the exposed original strands, according to base pairing rules, so that new complementary strands form.

4 The process of unzipping and joining new nucleotides continues along the whole length of the DNA molecule.

Each DNA molecule so formed is identical to the other and to the original DNA (and contains one strand of the original).

DNA replication is illustrated in Figure 19.

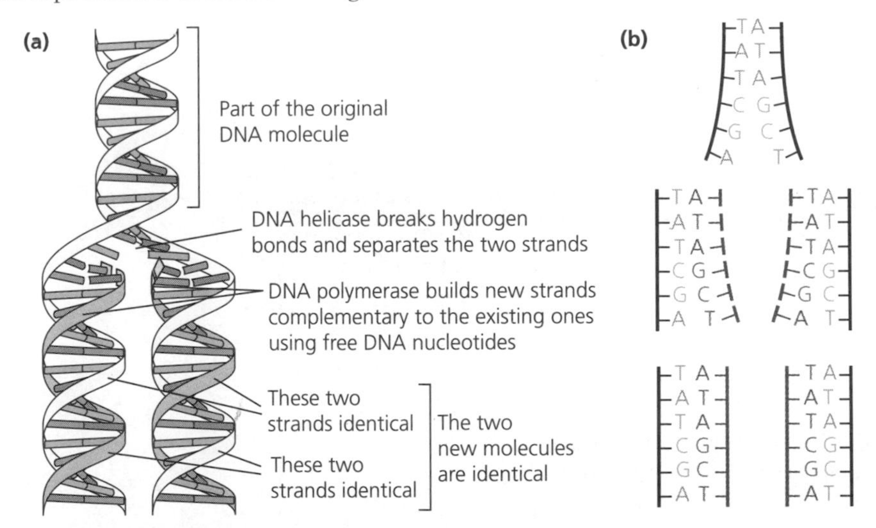

Figure 19 (a) DNA replication. (b) Semi-conservative replication of DNA

Exam tip

Learn the base pairing rules: A–T, C–G. It is common to ask students to write out the complementary sequence to a given DNA sequence.

Knowledge check 15

What is the sequence of bases on the complementary strand created by replication of a 'parental' strand with the sequence ATCTGTA?

Evidence for semi-conservative replication

Experimental evidence that DNA replicates semi-conservatively came from a classic experiment devised by Matthew Meselson and Franklin Stahl in 1958. They grew bacteria (*Escherichia coli*) in a medium in which nitrogen was supplied (in ammonium ions) in the form of the heavy, but non-radioactive, isotope ^{15}N. Consequently, the DNA of the bacteria became entirely heavy.

These bacteria were then transferred to a medium containing the normal (light) isotope, ^{14}N. Immediately before changing the medium and then at intervals corresponding to successive generations, samples of bacteria were removed and the DNA was extracted.

Analysis of the extracted DNA involved density-gradient centrifugation, a technique that separates molecules of different masses; heavier molecules are deposited at a lower level in the centrifuge tube. The results were as predicted by the semi-conservative hypothesis. Immediately before the change of medium, DNA occupied a single band corresponding to 'heavy' DNA. After one generation DNA was of 'intermediate' density, being concentrated in a single band a little higher up the tube. This was to be expected if all the DNA molecules consisted of one heavy strand and one light strand. After two generations there were two bands, with 50% of DNA 'intermediate' and 50% 'light'. The results and their interpretation are shown in Figure 20.

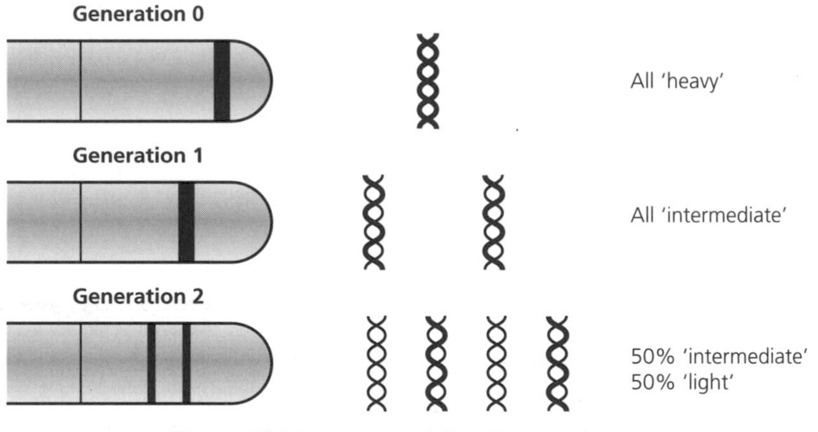

Figure 20 Meselson and Stahl's experiment

Knowledge check 16

Work out the proportions of light and intermediate DNA in the third generation. Use Figure 20 to help you.

Practical work

Use biochemical tests to detect the presence of carbohydrates and proteins:

- iodine test
- Benedict's test
- glucose-specific tests
- Biuret test

Carry out paper chromatography to identify amino acids:

- preparation, running and development of the chromatogram
- calculation of R_f values

Summary

- Water molecules are bipolar. They are attracted to each other and surround ions and molecules with charged groups, which are thus dissolved.
- A range of ions is required for the synthesis of many biological molecules.
- Carbohydrates consist of monosaccharides (single sugars), disaccharides and polysaccharides.
- Monosaccharides are classified according to the number of C atoms: pentose (5C) sugars, for example ribose and deoxyribose; hexose (6C) sugars, for example α-glucose, β-glucose and fructose.
- Disaccharides include: maltose formed from two α-glucose molecules; sucrose formed from an α-glucose and a fructose.
- The polysaccharides amylose and amylopectin (forming starch in plants) and glycogen (in animals) are stores of α-glucose molecules (which are used in respiration to release energy). The bonds between the α-glucose molecules are 1,4-glycosidic bonds, although amylopectin and glycogen have, in addition, 1,6-glycosidic bonds, which cause them to branch.
- Cellulose is a polysaccharide of β-glucose. Straight chains are formed that H-bond with each other to provide fibres of high tensile strength in plant cell walls.
- Triglycerides are lipids composed of three fatty acids joined to a glycerol by ester bonds.
- Fatty acids can be either saturated or unsaturated (in which case there is a double bond in the hydrocarbon chain). Fatty acids release energy in aerobic respiration.
- A phospholipid consists of glycerol bonded to two fatty acids and a phosphate. The glycerol-phosphate end is hydrophilic (water soluble) while the hydrocarbon chains are hydrophobic (water insoluble).
- Amino acids are joined by peptide bonds to form polypeptides. There are 20 different amino acids and, since any number can be joined in any order, an infinite variety of polypeptides is possible.
- Polypeptides form the basis of protein structure. Proteins have different levels of structure: primary structure (sequence of amino acids); secondary structure (α-helix or β-pleated sheet); tertiary structure (folding of polypeptide into a globular shape); and quaternary structure (with two or more polypeptide chains).
- The formation of glycosidic, ester and peptide bonds to build up macromolecules involves condensation reactions; their breakdown involves hydrolysis reactions.
- Brain tissue diseases, such as scrapie (in sheep), bovine spongiform encephalopathy (in cattle) and Creutzfeldt-Jacob disease (in humans), are caused by prion proteins, which are structurally altered forms (rich in β-sheets) of a normal, cell-surface glycoprotein.
- Nucleic acids are polymers of nucleotides.
- A nucleotide consists of a pentose sugar, a phosphate and an organic base.
- DNA consists of two anti-parallel strands, twisted to form a double helix, in which:
  - each strand contains nucleotides with the pentose sugar deoxyribose, though differing in the organic base — A, T, C or G
  - specific base pairing, with hydrogen bonds, occurs between the strands — A–T and C–G
  - phosphodiester bonds join the phosphate of one nucleotide to the deoxyribose of the neighbouring nucleotide along the 'sugar-phosphate backbone'
- RNA is a single-stranded chain of nucleotides with the pentose sugar ribose and each with one of four organic bases — A, C, G and U (instead of T).
- DNA replication involves the following stages:
  - DNA helicase catalyses the separation of the two strands.
  - Each strand forms a template, with DNA polymerase catalysing the binding of free deoxyribonucleotides according to base-pairing rules.
  - Two complete molecules of DNA are formed, identical to each other and to the original molecule of DNA.

Enzymes

The chemical reactions of an organism are collectively called **metabolism**. Metabolic reactions include:

- **catabolism** — 'breakdown' reactions
- **anabolism** — 'build-up' reactions

Enzymes are proteins that catalyse metabolic reactions — there is one type of enzyme for each reaction.

> **Knowledge check 17**
>
> Which of the following enzyme-controlled reactions in plant cells is anabolic and which is catabolic?
>
> **a** the formation of maltose from starch
>
> **b** the synthesis of starch from glucose-1-phosphate

The theory of enzyme action

In order to catalyse a reaction, the enzyme and substrate must first collide to form an **enzyme–substrate complex**. Catalysis then takes place on the enzyme surface, according to the equation below:

$$E + S \rightarrow ES \rightarrow EP \rightarrow E + P$$

enzyme + substrate → enzyme–substrate complex → enzyme–product complex → enzyme + product

The enzyme is unchanged at the end of the reaction.

Enzyme molecules have specific shapes determined by their tertiary structure (and, in some, their quaternary structure). The reaction takes place on a particular part of the enzyme molecule called the **active site**. In an anabolic reaction the substrate molecules are orientated in such a way on the active site as to allow bonding between them. In a catabolic reaction the formation of the active site round the substrate assists the breaking of a particular bond. In both cases the enzyme functions as a catalyst by effectively lowering the **activation energy** required for the reaction to take place. This reduction in activation energy is illustrated in Figure 21.

> **Knowledge check 18**
>
> What is the 'activation energy' of a reaction?

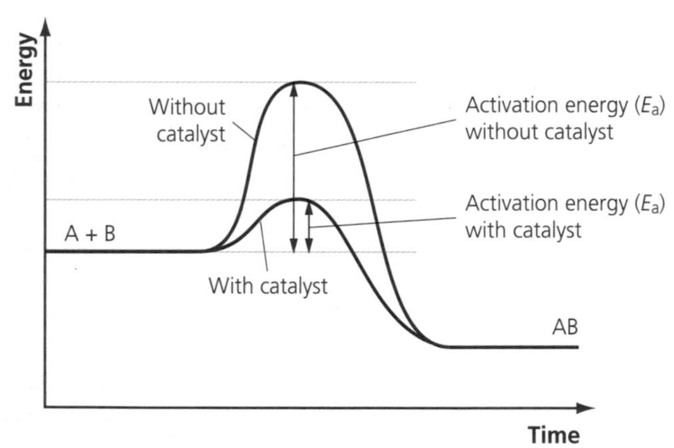

Figure 21 The effect of a catalyst on the activation energy of a reaction

Enzymes are **specific**. For any one type of reaction there is a particular enzyme required for catalysis — if there are 1000 different kinds of reaction in a cell, then the cell contains 1000 different enzymes.

There are two models to explain how enzymes work:

- the **lock-and-key** model (Figure 22)
- the **induced-fit** model (Figure 23)

The lock-and-key model of enzyme action proposes that the active site of an enzyme has a complementary shape (like a lock) into which the substrate molecule (the key) fits exactly, to form the enzyme–substrate complex. The induced-fit model suggests that, initially, the shape of the active site is not quite complementary to that of the substrate, but as the substrate begins to bind, the active site changes shape and 'moulds' itself around the substrate molecule (like a glove fitting round a hand). The induced-fit model is considered more useful because it better explains the way in which activation energy is reduced in catabolic reactions.

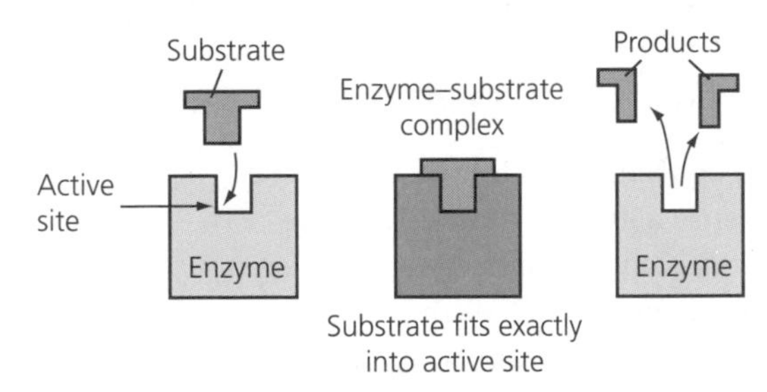

Figure 22 The lock-and-key model of enzyme action

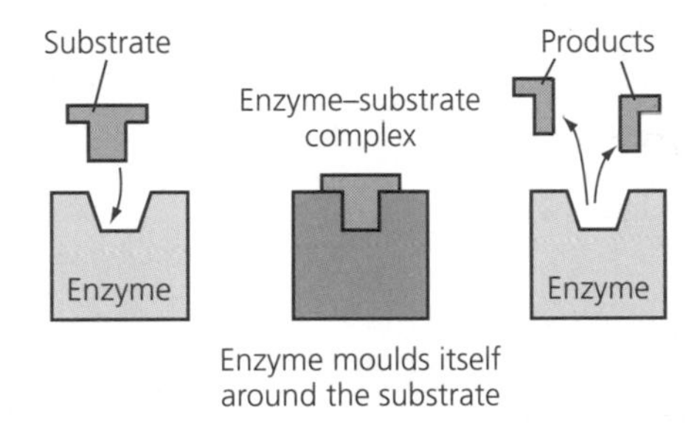

Figure 23 The induced-fit model of enzyme action

Properties of enzymes

A number of external factors influence the activity of enzymes and, therefore, the rate of biological reactions. These include substrate concentration, enzyme concentration, temperature and pH.

There are two key points to remember when explaining these influences:

- Enzyme molecules need to collide with substrate molecules, so factors that influence the chance of collision, such as substrate concentration and temperature, influence the rate of reaction.
- Enzymes are globular proteins with a precise tertiary structure, so factors that influence protein shape, such as high temperature and pH, influence the rate of reaction.

Knowledge check 19

Explain the specificity of enzymes.

Exam tip

You should understand how enzymes reduce activation energy in both anabolic and catabolic reactions. For anabolic reactions, the substrate molecules are held on the active site and orientated in such a position as to facilitate bonding between them. For a catabolic reaction, the binding of the substrate induces a conformational change in the shape of the active site that distorts the substrate molecule, so facilitating the breaking of a particular bond.

The effect of substrate concentration on enzyme activity

The influence of substrate concentration is shown in Figure 24.

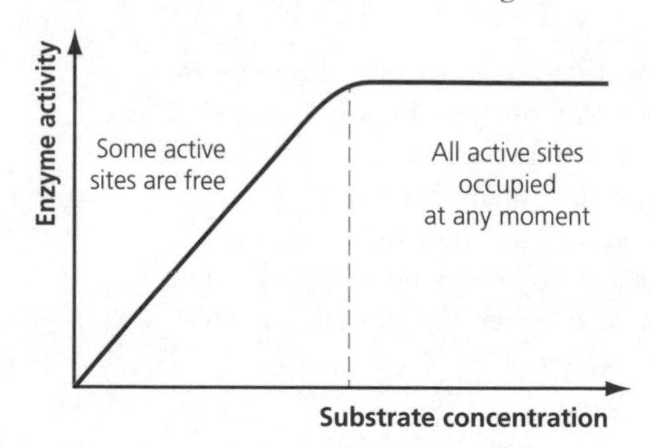

Figure 24 The effect of substrate concentration on enzyme activity

- At low substrate concentrations an increase in concentration increases enzyme activity — it is the **limiting factor**. This is because a greater concentration of substrate molecules increases the chances of collision with enzyme molecules. Therefore, more enzyme–substrate complexes are formed.
- At high substrate concentrations an increase in concentration does not cause a further increase in activity. This is because at high substrate concentrations the enzymes are fully employed and so, at any one moment, all the active sites are occupied.

The effect of enzyme concentration

The influence of enzyme concentration is shown in Figure 25.

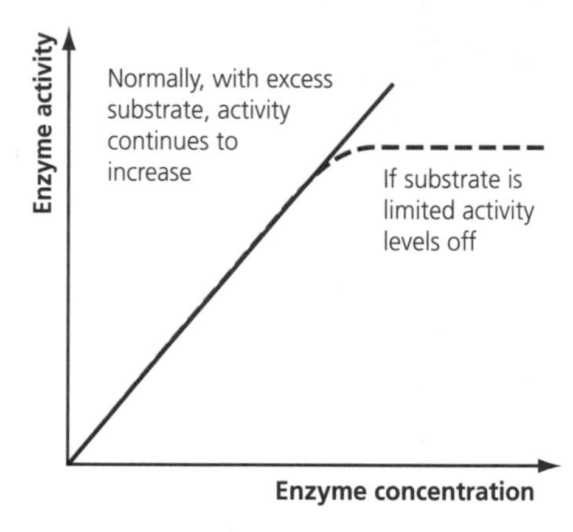

Figure 25 The effect of enzyme concentration on enzyme activity

An increase in enzyme concentration increases the rate of reaction. At high enzyme concentration activity may level off, but only if there is insufficient substrate. This is because an increase in the concentration of enzyme molecules increases the chance of successful collisions with substrate molecules. (There is normally only an incline phase since enzymes are used over-and-over again and so function efficiently at very low concentrations.)

Exam tip

If you were asked to describe the trends in Figure 24, for the incline part of the graph avoid simply saying that activity is 'faster' as if the *x*-axis represented 'time' — enzyme activity is increased as the concentration of substrate is increased. For the plateau part, don't be tempted to write 'there is no effect' — there is still a high level of activity, just no further increase.

Limiting factor The factor that determines the rate of a process. For example, at low substrate concentration it is the availability of substrate molecules which determines enzyme activity, and so this is the limiting factor.

Knowledge check 20

When there is an excess of enzyme present in an enzyme-catalysed reaction, explain the effects on the rate of reaction of increasing the concentration of:

a the substrate
b the enzyme

The effect of pH on enzyme activity

The influence of pH is shown in Figure 26.

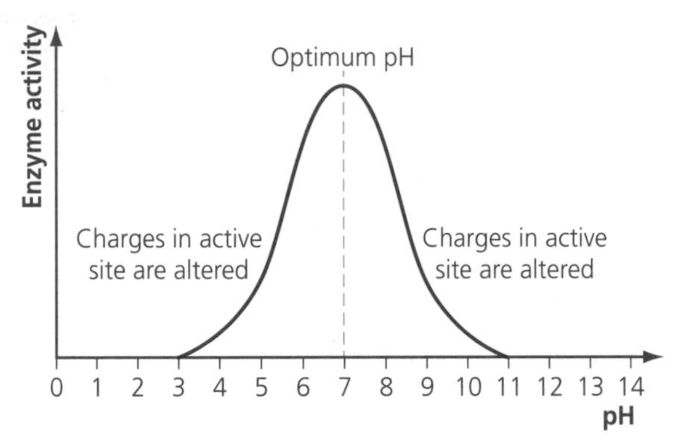

Figure 26 The effect of pH on enzyme activity

Changes from the optimum pH cause a decrease in enzyme activity. This is because the nature of the protein and, so, the active site of the enzyme are altered by changes in pH. In particular, ionic bonds in the tertiary structure are disrupted. pH is a measure of hydrogen ion (H^+) concentration and, as the concentration of hydrogen ions changes, the charges on the R-groups of amino acids are altered: at low pH a high concentration of H^+ causes negatively charged R-groups to lose their charge; at high pH a low concentration of H^+ causes positively charged R-groups to lose their charge (Figure 27). So at non-optimal pH the substrate attaches less readily to the enzyme and there is a specific pH at which the charges in the active site best facilitate the formation of an enzyme–substrate complex.

Exam tip

Minor changes in pH do not irreversibly denature enzymes. The ionic bonds disrupted can re-form if pH returns to optimum. However, at *extreme* changes of pH, denaturation can be permanent.

Optimum pH

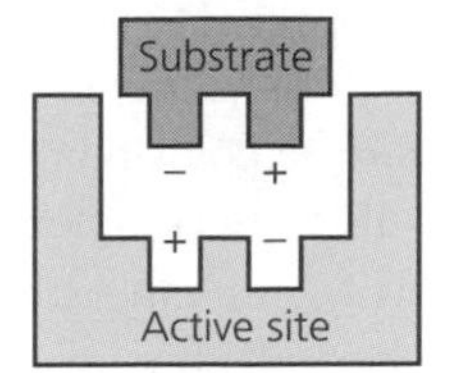

Charges on active site match those of substrate

Low pH

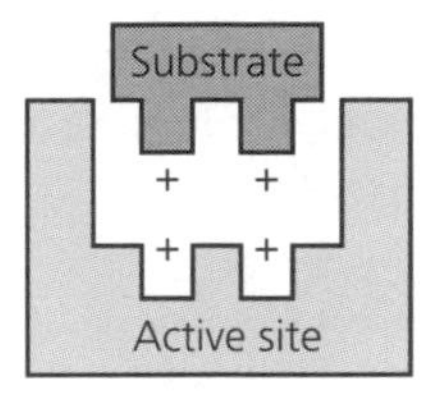

Charges on active site repel substrate

High pH

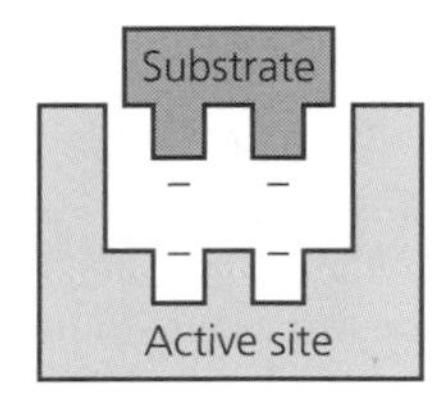

Charges on active site repel substrate

Figure 27 The effect of pH

The effect of temperature on enzyme activity

The influence of temperature is shown in Figure 28.

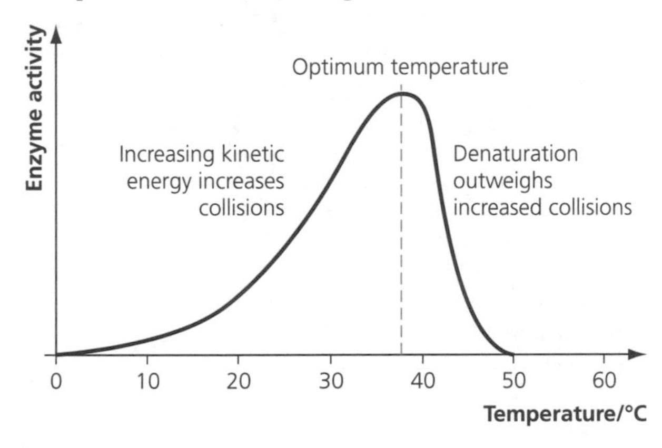

Figure 28 The effect of temperature on enzyme activity

Exam tip

If asked to explain the trends in the graph in Figure 28, you must show your understanding of how temperature affects kinetic energy, yet causes denaturation at high temperatures. It is a good idea to start answers to questions that ask you to 'explain' with 'Because...'.

- At low temperatures an increase in temperature causes an exponential increase in enzyme activity — typically, a 10°C rise in temperature doubles the rate of reaction. This is because an increase in temperature provides more kinetic energy for the collision of enzyme and substrate, so the rate of formation of enzyme–substrate complexes increases.
- At high temperatures (typically above 40°C) an increase in temperature causes a sharp decline in enzyme activity. This is because the bonds holding the tertiary structure of the enzyme molecules are broken and so the active site loses its complementary shape for substrate attachment — the enzyme is denatured.

Not all enzymes have an optimum temperature of around 40°C. Enzymes from organisms that live in very cold habitats have much lower optimum temperatures. Enzymes from thermophilic bacteria that live in hot springs are active at temperatures up to 90°C.

Exam tip

There is often confusion, when explaining denaturation, as to which bonds in the tertiary structure are broken by heat. Heat can break the weaker hydrogen and ionic bonds. It does not affect the disulfide bonds. It is the large number of disulfide bonds in heat-resistant enzymes (found in thermophilic bacteria) that provide increased thermostability.

Enzymes, cofactors and coenzymes

Some enzymes do not function effectively unless a non-protein **cofactor** is attached. Cofactors include metal ions, such as Mg^{2+}, and organic molecules (**coenzymes**) that are often derivatives of vitamins. Cofactors function either by influencing the shape of an enzyme (to its optimum for substrate attachment) or by participating in the enzymatic reaction (by attaching to one of the products for transfer to another enzyme). Some examples of cofactors and coenzymes are given in Table 6.

Table 6 Examples of cofactors and coenzymes

Enzyme	Cofactor	Role of enzyme
Carbonic anhydrase	Zinc ion (Zn^{2+})	Catalyses the combination of CO_2 with water to form carbonic acid in red blood cells, facilitating the transport of CO_2 in the blood
Cytochrome oxidase — a respiratory enzyme	Copper ion (Cu^{2+})	Combines electrons and hydrogen ions with oxygen in respiration
Enzyme	**Coenzyme**	**Role of enzyme and coenzyme**
Pyruvate decarboxylase — a respiratory enzyme	Coenzyme A	Pyruvate (3-carbon molecule) is broken down to acetate, which is 'picked up' by coenzyme A (forming acetyl CoA), and CO_2, which diffuses out of the cell
Succinate dehydrogenase — a respiratory enzyme	FAD (derived from vitamin B2, riboflavin)	Hydrogen is removed from succinate and 'picked up' by FAD (to form $FADH_2$)

Knowledge check 21

Why does the denaturation of an enzyme prevent its functioning?

Knowledge check 22

Chloride is a cofactor of amylase, the enzyme that catalyses the breakdown of starch to maltose molecules. What would happen to the production of maltose if amylase had to function in the absence of chloride ions?

Exam tip

The examples shown in Table 6 are provided only to illustrate the roles of cofactors and coenzymes. You do not need to learn these — most are involved in respiratory metabolism, which is covered in A2 Unit 2.

Enzyme inhibitors

Enzyme inhibitors are molecules that bind to enzymes and decrease their activity.

- A **competitive inhibitor** closely resembles the structure of the substrate and so competes for the active site, but does not remain there permanently (Figure 29).

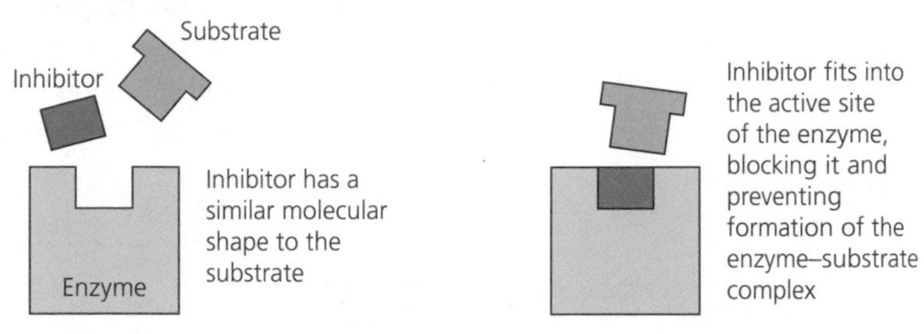

Figure 29 A competitive inhibitor competes with the substrate for the active site

- A **non-competitive inhibitor** does not resemble the substrate and may act in different ways:
  - Inhibitor molecules bind to a part of the enzyme away from the active site, altering the overall shape of the enzyme molecule, including the active site (Figure 30). The inhibitor may leave the enzyme so that the active site regains its catalytic shape.
  - Inhibitors bind to the enzyme molecule, leaving the enzyme permanently damaged. Some (e.g. cyanide, CN^-) combine irreversibly at the active site.

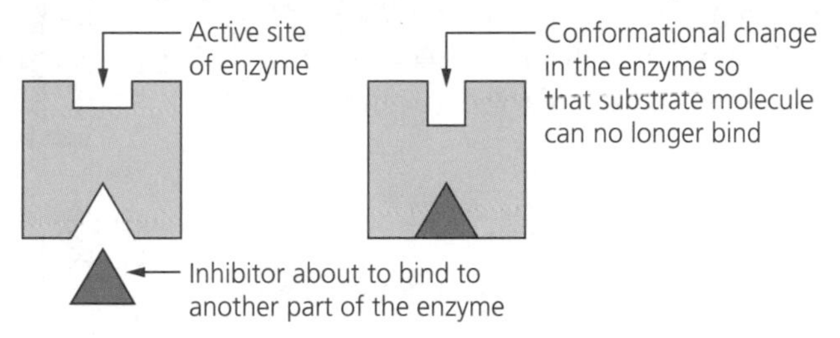

Figure 30 One type of non-competitive inhibitor

The effects of both types of inhibitor on enzyme activity at increasing substrate concentration are shown in Figure 31.

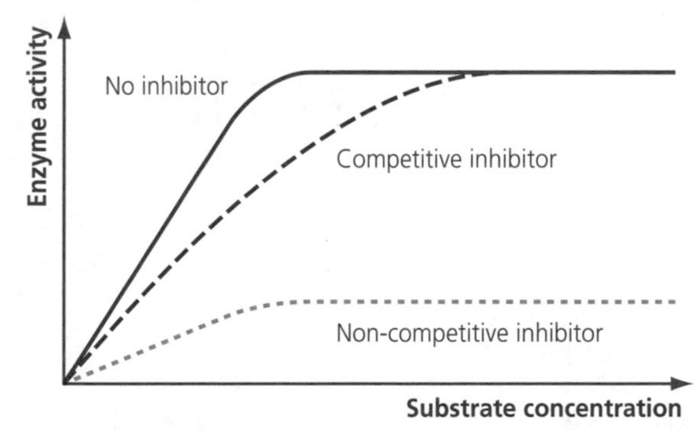

Figure 31 The effect of substrate concentration on enzyme activity in the presence of a competitive inhibitor and a non-competitive inhibitor

Exam tip

Inhibitors may be described as 'competitive' or 'non-competitive' and as 'reversible' or 'non-reversible (permanent)'. These terms can cause confusion. Remember that:

- competitive inhibitors are always reversible
- non-competitive inhibitors may be reversible or permanent
- non-reversible inhibitors are always non-competitive

Effect of a competitive inhibitor: the degree of inhibition depends on the relative concentration of both inhibitor and substrate because each is competing for a place on the active site. The more substrate there is available the more likely it is that a substrate molecule will find an active site. Therefore, if the substrate concentration is increased, the effect of the inhibitor is reduced.

Effect of a non-competitive inhibitor: the substrate and the inhibitor are not competing for the same site, so an increase in substrate concentration does not decrease the effect of the inhibitor.

Some examples of enzyme inhibitor are shown in Table 7.

Table 7 Examples of enzyme inhibitors

Enzyme	Inhibitor	Type
Succinate dehydrogenase — a respiratory enzyme	Malonate — similar in structure to succinate (the substrate)	Competitive inhibitor (reversible)
Cytochrome oxidase — a respiratory enzyme	Potassium cyanide, KCN — combines with the active site	Non-competitive inhibitor (and irreversible)

Knowledge check 23

Heavy metal ions, such as mercury (Hg^+), silver (Ag^+) and arsenic (As^+), alter the tertiary structure of a protein by breaking its disulfide bonds. What type of enzyme inhibitor would they be?

Immobilised enzymes

An enzyme is immobilised when it is confined so that it cannot move but still acts on its substrate. While there is a production cost in the immobilising process, enzyme immobilisation facilitates the use of continuous-flow column reactors (Figure 32), with a number of commercial advantages:

- Production can take place continuously.
- The product is enzyme-free, so purification costs are reduced.
- The enzyme remains separate from the reaction mixture and so can be easily re-used.
- Since the enzyme is supported, its stability is improved. This means that it remains active over a greater range of pH and temperatures (thermostability) than would be the case if the enzyme were in solution.

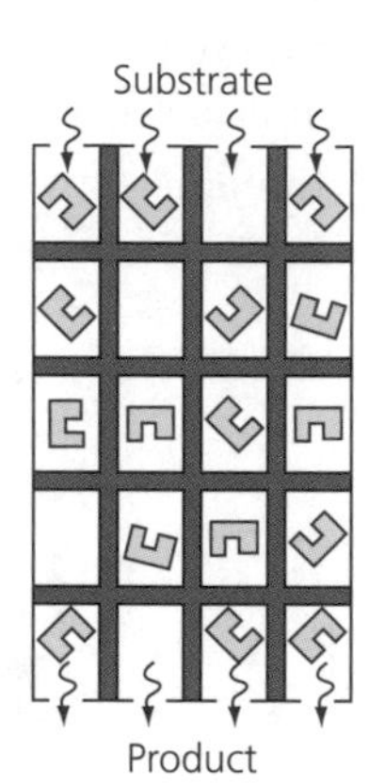

Figure 32 Immobilised enzymes in a continuous-flow column reactor

Different methods of immobilisation are illustrated in Figure 33.

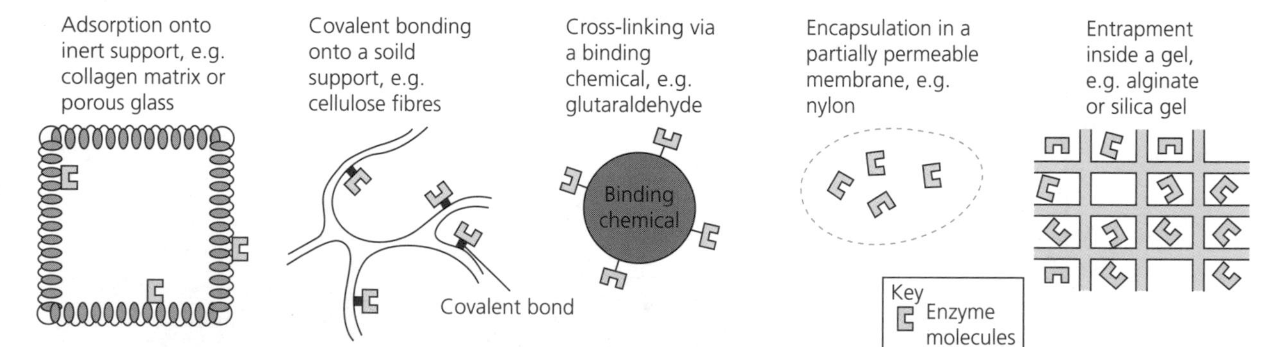

Figure 33 Different methods of enzyme immobilisation

Each method has its advantages and disadvantages (Table 8).

Table 8 Advantages and disadvantages of different types of immobilisation

Method	Advantages	Disadvantages
Adsorption	Easy to immobilise and so relatively cheap	Enzymes can easily be washed away; some active sites may be blocked by the adsorptive material
Covalent bonding	Enzymes not washed away; resistant to pH and temperature changes; most widely used method	Relatively expensive; some active sites may be blocked by support material or adversely altered in structure
Cross-linking	Enzymes securely bound, so not washed away; resistant to pH and temperature changes	Active sites may be blocked by binding chemical; significant distortion of some active sites during binding process
Encapsulation (enmeshment or membrane entrapment)	Enzymes not bound, so active sites are not blocked and activity is not adversely affected	Substrate has to diffuse through mesh; some enzymes may leak out through mesh
Entrapment (gel entrapment)	Enzymes cannot leak out; since enzymes not bound, active sites are not blocked	Resistance to substrates diffusing into a gel matrix and to products diffusing out

Exam tip

Be careful not to use the term 'thermostability' when referring to the increased stability of immobilised enzymes over a *range of pHs*.

Typically immobilisation reduces enzyme activity though it increases enzyme stability. For example, with the often-used covalent bonding method, some enzymes have blocked/altered active sites, so that the effective enzyme concentration is reduced. However, once immobilised on a support structure, the enzymes are less likely to change shape. The result is a different temperature profile (Figure 34).

Knowledge check 24

Using Figure 34 list three differences between the effects of temperature on the immobilised enzyme and on the free enzyme.

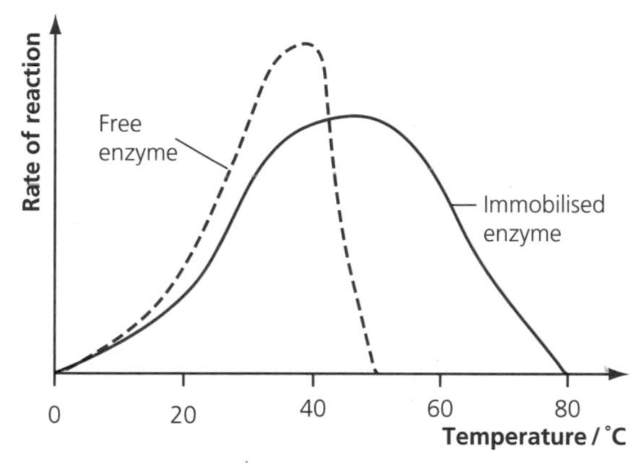

Figure 34 The effect of temperature on a free enzyme (in solution) and an immobilised enzyme

Diagnostic reagent strips as biosensors

Biosensors use immobilised enzymes to detect or monitor a particular molecule. The general principle is that the molecule reacts with a specific immobilised enzyme and the reaction causes a colour change or is converted into an electrical signal. They are used as **diagnostic** tools in industry, agriculture, science and medicine.

Biosensor A device that uses an immobilised enzyme (or antibody) to detect the presence of a particular chemical.

Diagnostic Allowing the identification of medical or other concerns.

Glucose test strips utilise the enzyme glucose oxidase, which specifically acts on glucose in the presence of oxygen to yield gluconic acid and hydrogen peroxide, with either product being measured according to the method of detection:

- **Glucose colour-change strips** (dipsticks), such as Clinistix or Dipstix, have, in addition to glucose oxidase, the enzyme peroxidase along with a colour dye immobilised on a paper pad (covered with a permeable membrane) at the tip of the strip. Peroxidase catalyses the breakdown of hydrogen peroxide (produced by glucose oxidase) to form water and oxygen; the oxygen released reacts with the colour dye, resulting in a *colour change*. The degree of colour change indicates the amount of glucose present. Test strips (Figure 35) are used to detect the presence of glucose in urine — a symptom of diabetes — or simply as a specific test for glucose.

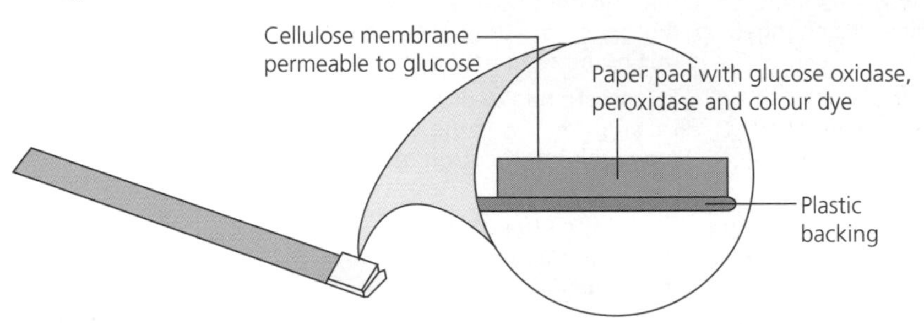

Figure 35 A glucose strip test

- **Glucose meters** measure the level of glucose in a blood droplet placed on a test strip, which is inserted in the meter. The amount of gluconic acid produced (by glucose oxidase) is converted into an *electrical signal*, which is then presented as a digital readout. Glucose meters are used by diabetics to monitor their blood glucose levels.

Some biosensors are based on enzyme inhibition. For example, heavy metal ions, as environmental pollutants, are detected through their inhibitory influence; the reduction in the biosensor's enzyme activity is a measure of the amount of pollution.

Knowledge check 25

Explain why glucose oxidase will not react with fructose.

Exam tip

The Clinistix (or Dipstix) test is not specified in the specification. Nevertheless, you should know examples of 'glucose-specific (monitoring) tests', which is specified, though you should not need to know the colour changes involved.

Exam tip

You should be aware that pregnancy test strips (another example of a biosensor) contain specific enzyme-linked antibodies that check for the presence of the hormone human chorionic gonadotropin (hCG) in urine.

Enzymes and medicine

Enzymes as biomarkers of disease

During injury or disease of an organ, the damaged tissue will release its enzymes, including those specific to the organ. Thus, measuring the level of particular enzymes in the body fluids into which they may be released can be used for diagnosis of disease or for monitoring the damage to a particular organ. Three examples of these enzyme biomarkers are given below:

- If heart muscle is injured, such as during a heart attack, the enzyme creatine phosphokinase-2 (CPK-2, the variant of CPK found in heart muscle) leaks out of the damaged cardiac tissue into the blood. Testing for its presence in **blood serum** can confirm whether a patient has suffered a heart attack.

Blood serum Blood plasma with fibrinogen removed to prevent clotting

- Certain white blood cells gather in the urinary tract if this becomes infected. There they release a specific variant of esterase. Testing for the presence of this esterase in urine indicates that white blood cells are active and is used to diagnose a urinary tract infection.
- During respiratory infection (e.g. pneumonia) certain white blood cells (called polymorphs) accumulate in the respiratory tract and release a form of the enzyme elastase. Testing for the presence of elastase in **sputum** allows early and effective treatment. This is important because elastase hydrolyses the structural protein elastin within the lung (elastin gives the lungs their elastic property), leading to reduced lung function and difficulty in breathing. Treatment would not only involve the use of antibiotics to tackle the infection, but the use of drugs that act as inhibitors of elastase, to prevent further lung damage.

Sputum Mucus coughed up from the respiratory tract, typically as a result of infection.

Enzyme inhibitors as therapeutic drugs

Many **therapeutic** drugs are enzyme inhibitors. The drug inactivates an enzyme associated with a particular medical condition though not with normal physiological activity. Examples of medical conditions where enzyme inhibitors are used as therapeutic drugs are provided in Table 9.

Therapeutic Relating to the healing of a disease or ailment.

Table 9 Examples of enzyme inhibitors used as therapeutic drugs

Medical condition	Inhibitor as drug	Target enzyme/reaction catalysed	Therapeutic action
Bacterial infection	Penicillin (an antibiotic)	Peptidoglycan transpeptidase — formation of cross-links in bacterial walls	The walls lose their strength, causing the bacteria to burst (due to osmotic intake of water)
Damaging effect of elastase during respiratory infection	α_1-antitrypsin (A1AT)	Elastase produced by white blood cells in respiratory tract	Prevents loss of elasticity in the lungs and reduction in lung function
Minor pain, caused by damaged tissue releasing prostaglandin	Aspirin	Prostaglandin-endoperoxide synthase 2, PTGS2 — synthesis of prostaglandin	Relief of symptoms such as minor pain and inflammation

Exam tip

The names of some enzymes and their inhibitors are only provided for illustration — you do not need to learn them Do not panic if you see names of enzymes or inhibitors within an exam question

Knowledge check 26

Why does penicillin not have an effect on human cells or viruses?

Knowledge check 27

What is a possible side effect of using high doses of aspirin for long periods?

The inhibitor used should have a high specificity for the target enzyme so that it does not bind to other enzymes and cause side effects. For example, to treat an infectious disease a drug is selected that targets an enzyme present in the pathogen but not in the host. However, side effects do occur. That of aspirin is well known — it blocks not only the enzyme PTGS2 but also PTGS1 which is needed to maintain a thick stomach lining, so long-term use of aspirin in high doses can lead to stomach bleeding and ulcers.

Practical work

Carry out experimental investigation of factors affecting enzyme activity:

- effect of temperature, pH, and substrate and enzyme concentration on enzyme activity
- demonstration of enzyme immobilisation
- use of a colorimeter to follow the course of a starch–amylase reaction (or other appropriate reaction).

Exam tip

A graphical skills question may ask you to 'plot the results, using an appropriate graphical technique'. You will have decisions to make: What type of graph is most appropriate? What is the most appropriate caption? Which is the independent variable?

Summary

- Each metabolic reaction involves a substrate being converted to a product and is catalysed by an enzyme.
- Catalysts speed up the reaction by lowering the activation energy required for the reaction to occur.
- Enzymes are globular proteins with a specific tertiary shape, with an active site.
- A substrate binds with the active site to form an enzyme–substrate complex. Products are formed at the active site. The products are released from the enzyme molecules, which are unaltered.
- There are two models of enzyme action:
  - The lock-and-key hypothesis explains enzyme specificity, due to the complementary shapes of the substrate and the enzyme's active site.
  - The induced-fit hypothesis suggests that binding of the substrate induces a change in enzyme structure which, through putting the substrate molecule under tension, explains why activation energy is lowered in catabolic reactions.
- Factors that increase the rate at which substrate and enzyme molecules might collide will increase the rate of reaction. They include temperature, substrate concentration and enzyme concentration.
- Factors that affect the tertiary structure of the enzyme will have an adverse effect on enzyme action by preventing binding of the substrate. They include high temperatures and pH changes away from the optimum.
- Cofactors are non-protein substances that are necessary for the actions of some enzymes. Cofactors can be ions or organic molecules (coenzymes).
- Inhibitors are substances that reduce the activity of enzymes:
  - Competitive inhibitors mimic the substrate and compete for the enzyme's active site; the extent of inhibition depends on the relative amounts of substrate and inhibitor.
  - Non-competitive inhibitors tend to stop enzyme activity and act in different ways.
- Enzymes can be immobilised, providing many advantages for their commercial use.
- Biosensors use immobilised enzymes to detect the presence of a particular molecule, for example glucose.
- Enzymes can be used as biomarkers of disease, where increased enzyme levels indicate infection or damage in an organ.
- Inhibitors of enzymes, associated with certain medical conditions, can be used as therapeutic drugs.

Cells and viruses

The cell is the structural unit of all living organisms. There are two categories of cell: **prokaryotic** and **eukaryotic**. Prokaryotic cells lack the degree of 'compartmentalisation' that eukaryotic cells possess — they lack a nucleus and other membrane-bound organelles. Bacteria are prokaryotes. Eukaryotes include animals, plants and fungi.

Viruses are not cells and are not themselves living. They have an intimate relationship with, and reproduce in, living cells.

Microscopy and cell ultrastructure

Two types of microscope are used in the study of cells: the **light microscope** and the **electron microscope**. Both **magnify** the fine structure of an object. However, **resolution** is even more important. This is the ability to discriminate fine detail so that two neighbouring points are seen as separate, rather than as a larger blur. The electron microscope has much greater resolving power than the light microscope because electrons have a shorter wavelength than light. The light microscope has its advantages, one of which is that living processes, such as mitosis, can be viewed. The interior of an electron microscope is a vacuum and so specimens must be dead. These two types of microscope are compared in Table 10.

Table 10 A comparison of the light microscope and the transmission electron microscope

Light microscope	Transmission electron microscope
Uses light: wavelength 450–700 nm	Uses electrons: wavelength 0.01 nm
Light refracted by glass lenses	Electron beams refracted by electromagnetic lenses
Low resolution: 200 nm	High resolution: 0.1 nm
Low magnification: ×1500 maximum	High magnification: ×1 000 000 maximum
Image formed on the retina of the eye or recorded on photographic film	Image formed on a fluorescent screen or recorded on photographic film
Limited in cellular detail that is revealed	Limited to dead specimens and by the likelihood of artefacts (deviations from the 'real' appearance as a result of the treatment of the specimen in preparation for microscopy)
Advantage: can be used to view living cells; also images are in colour	Advantage: produces high-resolution images of cells and organelles

The **scanning electron microscope** (SEM) is similar to the **transmission electron microscope** (TEM) except that the specimen is coated in a film of gold and electrons are reflected off the surface to create a three-dimensional effect image.

Exam tip

Remember that the main advantage of the electron microscope over the light (optical) microscope is *not* just that it provides a greater magnification but that it greatly increases the resolution — the ability to discern fine detail.

Exam tip

The advantage of the light microscope in being able to view living cells is important. For example, it allows the action of a phagocyte or the process of mitosis to be seen.

Exam tip

Students often get confused with the conversion of units. You should *measure* the object on the micrograph in millimetres (mm). Then a simple multiplication by 1000 presents the length in micrometres (μm).

Skills development

Numeracy skills: magnification

The specimen viewed using a microscope is called the **object**. It is increased in size by a certain **magnification** and presented in a micrograph as the **image**. The image, or a *scale bar* on the micrograph, can be measured so that knowing either the true (actual) size of the object, or the magnification, allows the other to be calculated. The magnification of a micrograph is calculated as:

$$\text{magnification} = \frac{\text{measured size of image}}{\text{true size}}$$

The true size of an object (e.g. an organelle) is calculated as:

$$\text{true size} = \frac{\text{measured size of image}}{\text{magnification}}$$

Of course, all *measurements must be in the same units*. The conversion factors are as follows:

$$\underset{(\text{m})}{\text{metres}} \underset{\times 1000}{\overset{\div 1000}{\rightleftarrows}} \underset{(\text{mm})}{\text{millimetres}} \underset{\times 1000}{\overset{\div 1000}{\rightleftarrows}} \underset{(\mu\text{m})}{\text{micrometres}} \underset{\times 1000}{\overset{\div 1000}{\rightleftarrows}} \underset{(\text{nm})}{\text{nanometres}}$$

Knowledge check 28

A scale bar, measuring 15 mm on a micrograph, represents a true length of 10 μm. Calculate the magnification.

Exam tip

The procedure for measuring the real (actual) size of an object is straightforward: measure the size of the image in mm, convert to μm and then divide by the magnification.

Cell fractionation and organelle isolation

To study the *function of cell organelles* they have first to be isolated. The technique for obtaining organelles is called **cell fractionation** and involves **homogenisation** and **centrifugation**. The procedure is summarised in Figure 36.

The fractions collected allow the function of different cell components to be investigated.

Knowledge check 29

Explain why the tissue is placed in a buffer solution.

Exam tip

You should understand that the isotonic solution stops the *organelles* bursting by osmosis. Some students say that it stops the cells from bursting, but of course they are deliberately burst by homogenisation.

(1) Chop up fresh liver tissue in ice-cold isotonic buffer solution

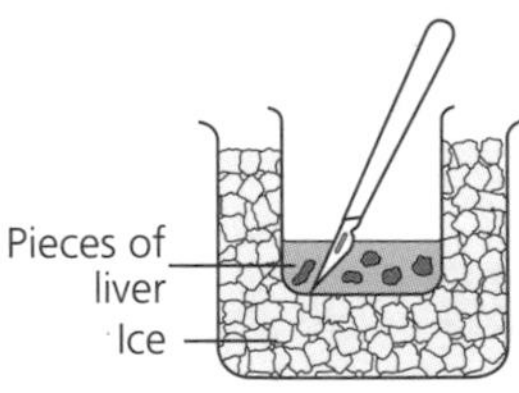

(2) Put tissue in blender or homogeniser to break open cells and release organelles

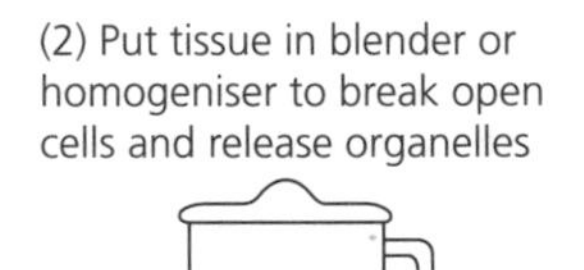

(3) Filter the mixture to remove debris

Ice

(4) Spin mixture in a centrifuge so that denser parts get thrown to the bottom forming a sediment (pellet)

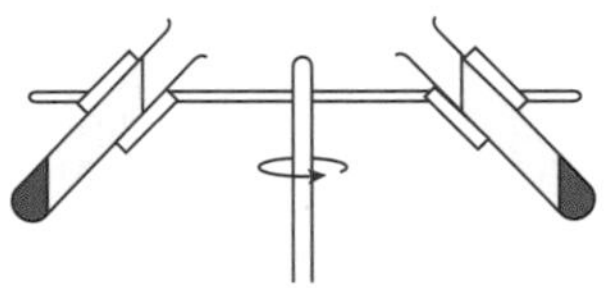

(5) The supernatant is poured into a tube, leaving a sediment (which contains nuclei)

(6) The supernatant can then be spun again at a faster speed to produce a sediment containing less dense organelles, e.g. mitochondria

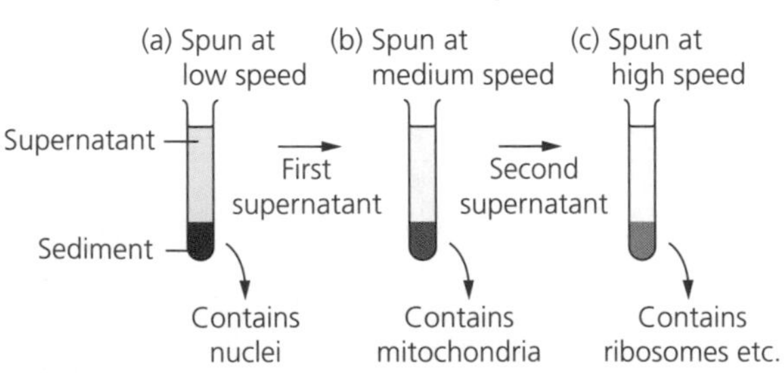

Figure 36 Cell fractionation involving homogenisation and centrifugation

Knowledge check 30

Explain how you might use this technique to obtain a sample of chloroplasts.

The eukaryotic cell

The appearance of a cell viewed via an electron microscope is its **ultrastructure**. Membrane-bound organelles, within which specific cellular functions take place, are evident. The ultrastructure of a generalised animal cell is shown in Figure 37.

Eukaryotes also include plants and fungi. Their cells have some distinctive features (Figure 38 — cells not drawn to scale).

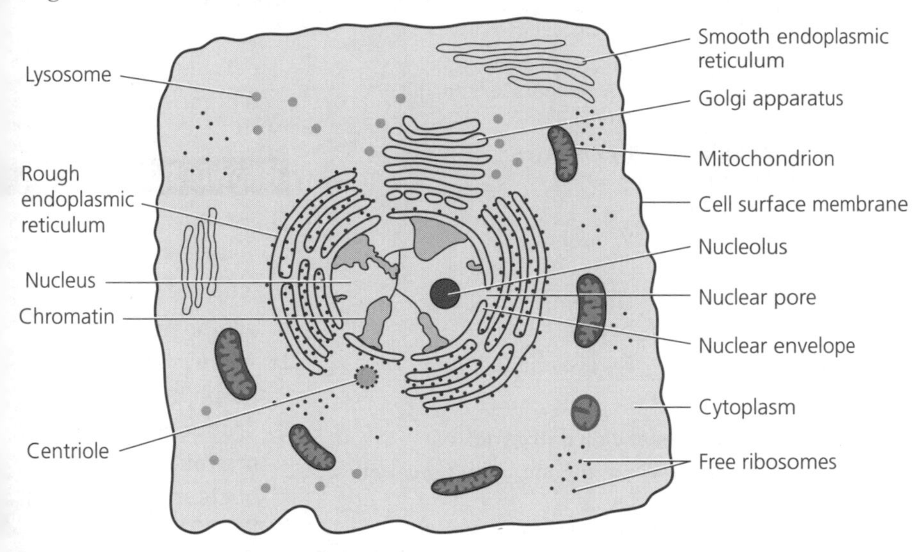

Figure 37 The ultrastructure of an animal cell

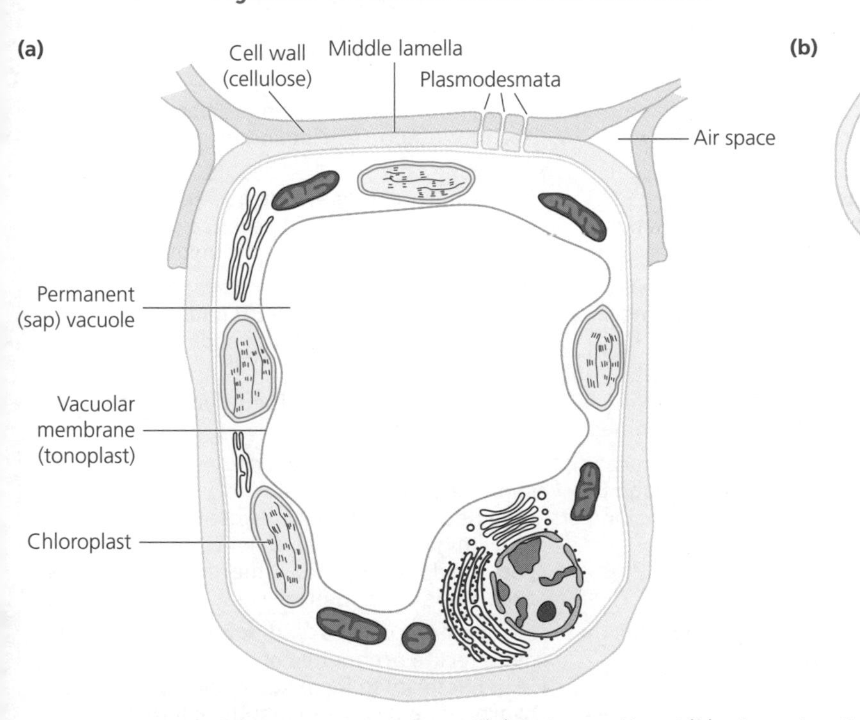

Figure 38 Distinctive features of (a) a plant cell and (b) a fungal cell

Exam tip

It is easy to press the wrong button on a calculator. Remember to check that your answer makes sense — a mitochondrion cannot be a metre in length!

Exam tip

You must show working and units when answering questions involving calculations. Then, even if the final answer is incorrect, you may receive some credit for showing correct operations within the calculation.

Both plant and fungal cells have cell walls, though made of different materials. The cell wall prevents the cell bursting when in dilute solution.

Animal, plant and fungal cells are compared in Table 11.

Table 11 A comparison of animal, plant and fungal cells

Animal cell	Plant cell	Fungal cell
No cell wall	Cellulose cell wall	Chitin cell wall
No chloroplasts	Chloroplasts	No chloroplasts
Glycogen granules (carbohydrate (energy) store)	Starch grains (carbohydrate (energy) store)	Glycogen granules (carbohydrate (energy) store)
Lysosomes	No lysosomes (though hydrolytic enzymes may be present in the sap vacuole)	Lysosomes
No permanent vacuole	Large permanent, central vacuole	Vacuole
Centrioles	No centrioles (except mosses and ferns)	No centrioles (except one group)
No plasmodesmata	Plasmodesmata	No plasmodesmata

Knowledge check 31

List:

a three features present in a plant cell but not found in an animal cell

b three features present in a fungal cell but not found in a plant cell

Knowledge check 32

Name the organelles with the following functions:

a synthesis of lipids

b ATP synthesis in aerobic respiration

c production of spindle microtubules during nuclear division

The structures that perform particular functions in a cell are called **organelles** (Table 12). Some of these are surrounded by membranes — membrane-bound organelles.

Table 12 The structure and function of eukaryotic cell organelles

Organelle	Structure	Function
Nucleus	Largest organelle (10–30 μm) enclosed within an envelope (double membrane); contains chromatin, consisting of DNA wound round beads of histone proteins; perforated envelope (possesses pores); contains one or several nucleoli (1–2 μm)	DNA codes for the synthesis of polypeptides in the cytoplasm; pores in the envelope allow large molecules in (e.g. enzymes) and out (e.g. RNA); nucleolus synthesises ribosomal RNA and manufactures ribosomes
Ribosomes	Small bodies (20–25 nm) of protein and RNA either free in the cytoplasm or attached to rough endoplasmic reticulum	Site of polypeptide synthesis; free ribosomes (i.e. not attached to ER) produce proteins that will function within the cytoplasm
Rough endoplasmic reticulum (rough ER)	Membrane system of flattened sacs, continuous with the outer nuclear membrane and covered with ribosomes	Polypeptides made on the ribosomes accumulate in the rough ER and are passed on, in vesicles, to the Golgi apparatus
Smooth ER	Separate membrane system of interconnecting tubules (lacking ribosomes)	Synthesis of lipids and their distribution throughout the cell
Golgi apparatus	A stack of membrane-bound sacs (cisternae); forming face has vesicles from the rough ER joining it; mature face has vesicles pinching off	Dynamic structure in which polypeptides are combined (forming quaternary proteins) or modified (e.g. with carbohydrate attached to form glycoproteins); finished protein is packaged into vesicles either for secretion by exocytosis or for delivery elsewhere in the cell
Lysosomes	Vesicles produced by the Golgi apparatus that contain hydrolytic enzymes	Lysosomes combine with membrane-bound degenerate organelles or ingested particles (e.g. bacteria) to form secondary lysosomes; hydrolytic enzymes digest the contents (Figure 39)

→

Organelle	Structure	Function
Mitochondria (singular: mitochondrion)	Sausage-shaped (1 µm wide and up to 10 µm long); surrounded by an envelope, the inner membrane of which is folded to form cristae; fluid-filled matrix; several to thousands per cell	Synthesis of ATP by aerobic respiration
Chloroplasts	Ovoid (2–10 µm in diameter); surrounded by an envelope; elaborate internal membrane system of lamellae with thylakoids stacked into grana; contain lipid droplets and starch grains; found in plant cells	Site of photosynthesis; chlorophyll molecules are attached to the lamellae
Vesicles and vacuoles	Bound by a single membrane; vesicles are much smaller than vacuoles; vacuoles are permanent in plant and fungal cells; membrane of the sap vacuole in plant cells is called the tonoplast	Vesicles may be used for storage and transport of substances (e.g. transport to and from the cell-surface membrane or within the cytoplasm); vacuoles are for storage of water and ions
Microtubules	Tubular (25 nm in diameter); formed from the protein tubulin; occur within centrioles (as nine triplets of microtubules in a circular arrangement) and throughout the cytoplasm; animal and fungal cells contain a pair of centrioles	Centrioles form the spindle fibres during cell division of animal and fungal cells; microtubules also form part of the cytoskeleton and allow movement of cell organelles
Microvilli	Finger-like folds of the cell-surface membrane	Increase the surface area for absorption of molecules and ions
Plasmodesmata (singular: plasmodesma)	Strands of cytoplasm between neighbouring plant cells that pass through pores in the walls	Facilitate transport of materials between adjacent cells in plants

Exam tip

Ribosomes should be described as the site of polypeptides synthesis rather than of protein synthesis. This is because polypeptides may be transported to the Golgi body where the finished protein is formed (quaternary protein or conjugated protein).

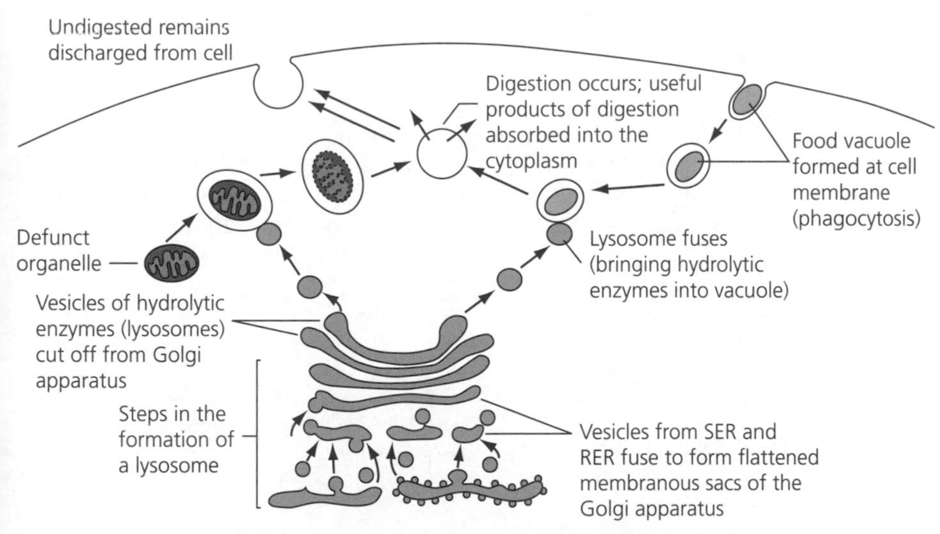

Figure 39 The role of lysosomes

Knowledge check 33

Name the organelle in each case with the following structures:

a cisternae
b cristae
c thylakoids

Knowledge check 34

What cellular processes would be taking place in a cell with:

a much rough endoplasmic reticulum
b a prominent Golgi body
c a cell-surface membrane convoluted to form many microvilli?

Exam tip

Students often confuse secretory vesicles and lysosomes. The Golgi apparatus produces *both*. However, their roles are quite distinct. Secretory vesicles are moved to the cell-surface membrane and their contents are exocytosed (p. 46). Lysosomes remain in the cell where they are involved in intracellular digestion (Figure 39).

Knowledge check 35

How are the functions of the following organelles linked?

a nucleolus and ribosomes

b rough endoplasmic reticulum, Golgi body and vesicles

The prokaryotic cell

Prokaryotic cells are particularly small (1–10 μm). They lack a nucleus and other organelles bound by membranes (Figure 40).

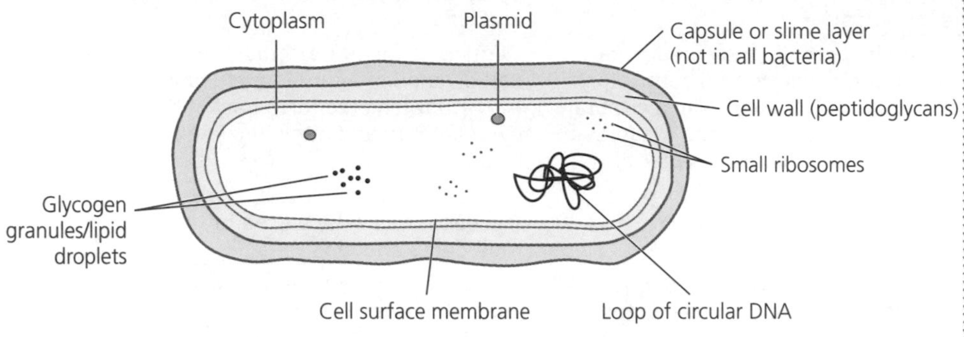

Figure 40 A generalised prokaryotic cell

Prokaryotic and eukaryotic cells are compared in Table 13.

Table 13 Comparison of prokaryotic and eukaryotic cells

Prokaryotic cell	Eukaryotic cell
Small cells — 1–10 μm	Large cells — 10–100 μm
No membrane-bound organelles	Nucleus, mitochondria, endoplasmic reticulum, Golgi apparatus and chloroplasts (in plants) present
Small ribosomes — 20 nm in diameter (70 S)	Large ribosomes — 25 nm in diameter (80 S)
Single circular DNA molecule, without associated protein; the region in the cytoplasm containing the DNA is called the nucleoid	DNA as several linear molecules associated with protein (histones) to form chromatin; these are contained within a membrane-bound nucleus
Plasmids (small circular pieces of DNA outside the main DNA molecule) usually present	No plasmids
Peptidoglycan cell wall	Cellulose cell wall present in plant cells; and chitin cell wall in fungal cells
No microtubules	Microtubules present and organised into centrioles in animal and fungal cells
Slimy outer capsule may be present	No capsule

Exam tip

Be prepared to make a comparison between a prokaryotic cell and a eukaryotic cell. Make sure that any differences noted are sufficiently distinct. For example, the examiners might regard 'mitochondria absent in prokaryotes' and 'endoplasmic reticulum absent in prokaryotes' as a single point — i.e. 'membrane-bound organelles absent in prokaryotes'.

Viruses

Viruses lack cytoplasm and are not cells. They consist of a nucleic acid core surrounded by a protein coat, called a **capsid**. The nucleic acid ultimately acts as a coding device for the production of new viral particles. There are a number of types (Figure 41).

Knowledge check 36

What are the two main components of a virus particle?

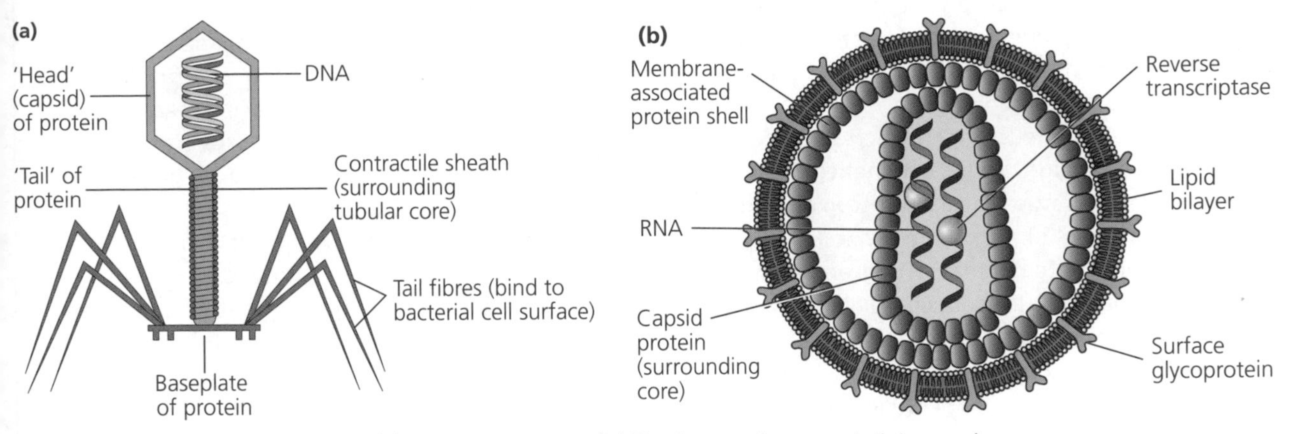

Figure 41 (a) A bacteriophage. (b) The human immunodeficiency virus

Bacteriophages

Bacteriophages (phages) consist of a core of double-stranded DNA bounded by a protein coat. Phages invade bacteria and the phage DNA codes for the production of new viral proteins (to make new coats). The phage DNA replicates to form many copies, which are then packaged in the new protein coats. Eventually the bacterial cells burst to release many new phages.

Knowledge check 37

List three differences between viruses and living cells.

Human immunodeficiency virus (HIV)

HIV consists of a core of RNA bounded by a protein coat and a lipid bilayer containing glycoproteins. It belongs to a group of viruses containing RNA that are known as **retroviruses**. They contain the enzyme **reverse transcriptase**, which uses the RNA as a template to produce single-stranded DNA; double-stranded DNA is then created with DNA polymerase activity. This viral DNA is integrated into the host DNA where it ensures that viral protein and new copies of RNA are made. HIV invades a type of lymphocyte (helper T-cell) and so may weaken the immune system, thereby causing AIDS — acquired immune deficiency syndrome.

Knowledge check 38

A section of the RNA in the HIV has the base sequence AUAUGUACTC. What is the corresponding base sequence on the DNA strand produced by the enzyme reverse transcriptase?

Practical work

Examine photomicrographs and electron micrographs (TEM/SEM):

- Recognise cell structures from photomicrographs and electron micrographs (TEM/SEM).
- Calculate true size (in μm) and magnification, including the use of scale bars.
- Draw individual cells or cell sections.

Examine fresh tissues:

- Stain tissues to aid observation when using a microscope (e.g. using iodine or methylene blue).
- Use a graticule and stage micrometer to measure cell length.

Exam tip

You should to be able to identify organelles in micrographs. Past papers containing micrographs are available on the CCEA website to allow ample practice.

Summary

- Electron microscopes produce images with greater magnification and, more importantly, greater resolution than light microscopes.
- Cell fractionation, involving homogenisation and centrifugation, is used to separate cell components (e.g. organelles) so that their function can be studied.
- Prokaryotic cells (bacteria) lack a nucleus and other membrane-bound organelles. They have smaller ribosomes, a circular loop of naked DNA and a cell wall of peptidoglycan.
- Eukaryotic cells (animals, fungi and plants) are compartmentalised, containing numerous organelles, many of which are membrane-bound, specialising in different metabolic activities.
- The nucleus encloses the genetic material (DNA), contained within chromatin, and is bounded by a perforated nuclear envelope.
- The endoplasmic reticulum consists of sheets of membranes. Rough ER is covered in ribosomes. Smooth ER is the site of lipid synthesis.
- Ribosomes, found on the rough ER or free in the cytoplasm, are the site of polypeptide synthesis.
- The Golgi apparatus consists of a stack of membrane-bound cisternae. It modifies proteins and produces secretory vesicles and lysosomes.
- Lysosomes contain lysozymes and are responsible for intracellular digestion.
- Mitochondria are enclosed by a double membrane, the inner of which is folded to form cristae, and is the site of ATP production by aerobic respiration.
- Plant cells differ from animal cells in having a cellulose cell wall (though with plasmodesmata connecting cells) and often having a sap vacuole. They might also have chloroplasts and lack centrioles, and they possess starch grains rather than glycogen granules.
- Centrioles contain microtubules and are a focus for spindle formation in animal cells.
- Chloroplasts are bounded by an envelope, have internal membranes (lamellae) with thylakoids stacked into grana, and contain the fluid stroma. They are the site of photosynthesis.
- Fungal cells differ in a number of ways from both animal and plant cells.
- Viruses are not cells and only become active within a living host cell. They are able to instruct the cell's metabolism to make new viral particles.
- Viruses consist of a core of nucleic acid and a protein coat (capsid).
  - Bacteriophages (host: bacteria) contain DNA.
  - HIV (host: human helper T-cells) contains RNA and the enzyme reverse transcriptase (producing viral DNA).

Membrane structure and function

The membranes within cells and surrounding them (the cell-surface or plasma membrane) consist of a phospholipid bilayer with associated proteins.

The structure of the cell-surface membrane

Phospholipid bilayers form spontaneously: the hydrophobic 'tails' lie innermost (shielded from the aqueous environment); the hydrophilic 'heads' face outermost (in contact with cytoplasm or extracellular fluid). **Cholesterol** molecules occur among the hydrocarbon tails (particularly in animal cells). **Proteins** float in the bilayer: **intrinsic proteins** span both layers (transmembrane) or are embedded into one layer; **extrinsic proteins** are attached to the membrane surface. The phospholipids in the cell membrane are constantly moving while the proteins are scattered among them, so that the structure proposed is called the **fluid-mosaic model**.

Exam tip

You should understand that proteins in contact with the hydrophobic centre of the bilayer consist of amino acids with hydrophobic (non-polar) R-groups.

Knowledge check 39

How does an intrinsic protein differ from an extrinsic protein?

The membrane also contains **polysaccharides** bound either to the proteins (**glycoproteins**) or to lipids (**glycolipids**). The polysaccharides are on the outer face only, where they form a fringe called the **glycocalyx**.

The fluid-mosaic structure of membranes is illustrated in Figure 42.

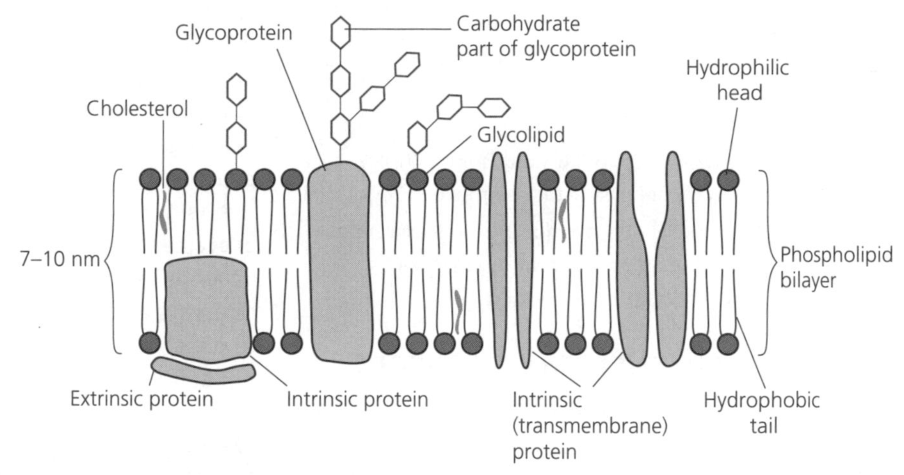

Figure 42 The structure of the cell-surface membrane

Knowledge check 40

Explain why cell-surface membranes are not all the same.

Membrane fluidity

A number of factors influence the fluidity of the membrane:

- The more phospholipids with **unsaturated hydrocarbon chains** there are, the more fluid is the membrane. The 'kinks' in the unsaturated hydrocarbon tails prevent them from packing close together, so more movement is possible.
- Phospholipids with **longer hydrocarbon chains** decrease the fluidity of the membrane, since attractive forces among the tails will be greater.
- Fluidity is influenced by **temperature**. The membrane is more fluid at high temperature and less fluid at low temperature as the phospholipid bilayer 'freezes' into a gel (or solid-like) state.
- **Cholesterol** acts as a temperature stability buffer. At high temperature cholesterol provides additional binding forces and so decreases membrane fluidity. At low temperature cholesterol keeps the membrane in a fluid state by preventing the phospholipids from packing too close together and 'freezing'.

Cell recognition and cell receptors

Glycoproteins and glycolipids have important roles in **cell-to-cell recognition** and as **receptors** for chemical signals. The glycocalyx allows cells to recognise each other and, therefore, group together to form tissues. Glycoprotein receptors and signalling molecules (e.g. hormones) fit together because they have complementary shapes.

Knowledge check 41

Suggest why liver and muscle cells are particularly responsive to the hormone insulin.

Membrane enzymes

Many of the proteins in the membrane are enzymes. The membrane provides the attached enzymes with improved stability (as with immobilised enzymes).

Movement of substances in and out of the cell

The cell-surface membrane acts as a barrier between the cytoplasm and the extracellular fluid, though exchange of substances does take place across it. The route taken to cross the membrane and the mode of transport depend on a number of factors:

- The **size** of the molecule — very small molecules can slip between the phospholipid molecules; large particles can only move in or out by cytosis (bulk transport).
- The **polarity** or **non-polarity** of the substance — non-polar (and lipid-soluble) molecules move through the phospholipid bilayer; polar substances move via proteins.
- The **concentration** of the substance either side of the membrane — substances move from high to low concentration by diffusion. If movement against the concentration gradient is required, then active transport is needed.

Exam tip

Remember that polar or water-soluble substances cannot pass through the phospholipid bilayer because of its hydrophobic centre and so must pass via intrinsic (transmembrane) proteins. Non-polar or fat-soluble substances are able to pass through the phospholipid bilayer.

Passive transport across the cell-surface membrane

Passive movement through the membrane occurs down a concentration gradient and does not require energy expenditure from ATP.

Knowledge check 42

Suggest two properties that a drug should possess if it is to enter a cell rapidly.

Diffusion

Non-polar molecules, such as lipid-soluble vitamins (e.g. vitamins A and D), steroid hormones, the respiratory gases oxygen and carbon dioxide, and very small polar molecules such as water and urea, move through the phospholipid bilayer down their concentration gradients.

Exam tip

It is not accurate to say that diffusion (passive movement) does not require *energy*. It relies on the kinetic energy of the substances in solution. It does not require energy from ATP (respiration).

Facilitated diffusion

Ions and molecules with charged groups, such as glucose and amino acids, are polarised and cannot move through the phospholipid bilayer. For these substances, movement may be facilitated (i.e. made easier) by particular membrane proteins, though no metabolic energy is used and movement is down the concentration gradient — hence the term **facilitated diffusion**.

Channel proteins have a hydrophilic core through which ions and water may diffuse. These channels have a specific shape to allow only one type of ion through. In many cases the channels can be opened and closed (called gated channels) to regulate flow. While some water molecules diffuse through the phospholipid bilayer (since they are so small), most water diffuses through specific water channel proteins called **aquaporins**.

Carrier proteins may selectively (specifically) transport medium-sized molecules such as glucose and amino acids. The molecule binds to a site on the protein, which changes shape to bring the molecule through the membrane.

Knowledge check 43

Distinguish between channel proteins and carrier proteins in the facilitated diffusion of water-soluble substances.

Skills development

Interpretation from graphs

You need to develop your ability to interpret information provided in a graph, often one that is unfamiliar to you. When presented with a graph you must spend some time studying it: read the axes carefully, remembering that the *y*-axis variable is dependent on the variable on the *x*-axis (the independent variable). The interpretation will involve you being asked to:

- '*describe* the trends in the graph' — this means 'turn the pattern shown into words'
- '*explain* the trends shown' — this means 'give biological reasons for what is happening'

Ensure that you give a full answer: sometimes students merely describe the trends and fail to explain them; or describe a single trend when more than one is evident.

The graphs shown in Figure 43 show the effect of difference in concentration across a membrane on the rate of (a) diffusion and (b) facilitated diffusion. Use these to practise your interpretation skills, applying your understanding of these two routes for membrane transport.

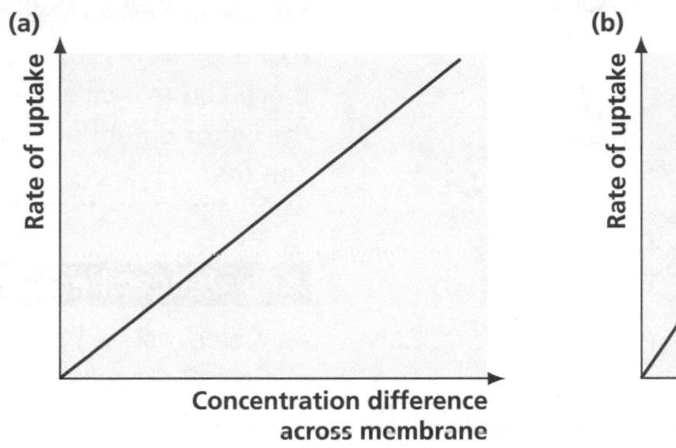

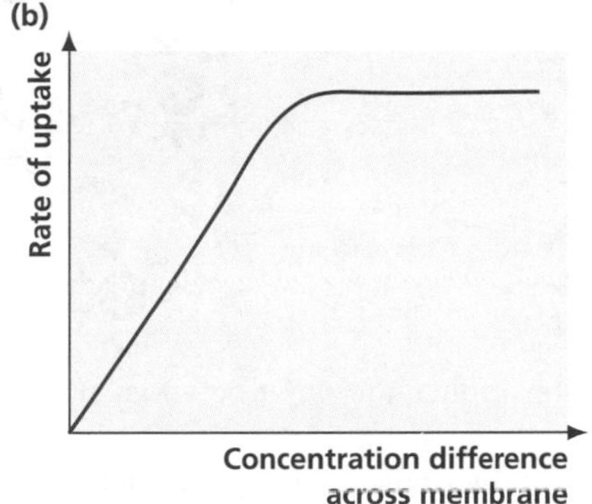

Figure 43 The effect of difference in concentration across a membrane on the rate of (a) diffusion and (b) facilitated diffusion

Descriptions and explanations are given in Table 14 for comparison with your own attempts.

Table 14 Descriptions and explanations of the trends in the graphs in Figure 43

	Diffusion	Facilitated diffusion
Description	As the concentration difference between the inside and outside of the cell increases the rate of uptake by diffusion increases	As the concentration difference increases so does the rate of uptake. However (in a second trend), at very high concentration differences the rate of facilitated diffusion reaches a maximum and levels off
Explanation	With a greater net concentration outside the membrane then the greater the chance of particles colliding with the phospholipid bilayer to gain entry.	The increase phase can also be explained by increased collisions with membrane proteins. The levelling off is due to the carrier or channel proteins in the membrane that are being used for facilitated diffusion becoming saturated

Osmosis

Osmosis is the diffusion of water across a differentially permeable membrane (water moves through more easily than solutes). Water moves from an area of higher **water potential** (Ψ) to an area of lower water potential. Pure water (at standard temperature and pressure) is defined as having a water potential of zero. The addition of **solutes** to water lowers its potential (makes it more negative), just as an increase in **pressure** increases its potential (makes it more positive).

water potential of a cell (Ψ_{cell}) = solute potential (Ψ_s) + pressure potential (Ψ_p)

The water potential is a measure of the free energy of the water molecules in a system. The water potential of pure water is zero because all the molecules are 'free'. In solutions, some of the water molecules form shells around the solutes and are, therefore, no longer free (Figure 44). This lowers the water potential — it becomes negative. Cytoplasm contains solutes, so it has a negative water potential. Therefore, when cells are placed in water, water moves into the cells by osmosis.

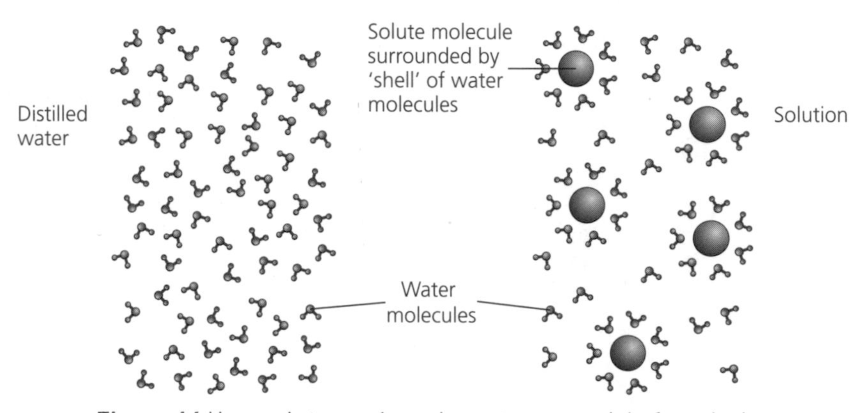

Figure 44 How solutes reduce the water potential of a solution

When placed in dilute (hypotonic) solutions, animal cells, such as red blood cells, take up water and swell until they burst (**lyse**). However, the cells of prokaryotes, fungi and plants have rigid walls that prevent them from bursting. The pressure created by the swelling cell increases to a point when water can no longer enter. A swollen plant cell is said to be **turgid**.

When red blood cells are placed in a concentrated (hypertonic) solution, the cells lose water by osmosis, shrink and become **crenated**. In hypertonic solutions, cells with walls also shrink — the cell wall cannot protect them from water loss by osmosis. As they shrink, the cells lose contact with their cell walls. In plant cells this is known as **plasmolysis** and the point at which the cytoplasm just begins to lose contact with the cell wall is called **incipient plasmolysis**.

Exam tip

Students often lose marks because they do not understand that all water potential (and solute potential) values are negative. The highest water potential is zero. Therefore the lower the water potential, the more negative it becomes.

Exam tip

Always describe the movement of water in terms of water potential (and *never* in terms of water concentration).

Knowledge check 44

Explain why dissolving more solute decreases the water potential of a solution.

Knowledge check 45

As water moves into a cell placed in a hypotonic (more watery) solution, what happens to the water potential of the cell?

Exam tip

You will need to learn the effect of immersing plant and animal cells in either hypotonic or hypertonic solutions.

Water potential, the solute potential of the cell contents and the pressure potential of a plant cell all change as the cell takes up or loses water osmotically (Figure 45).

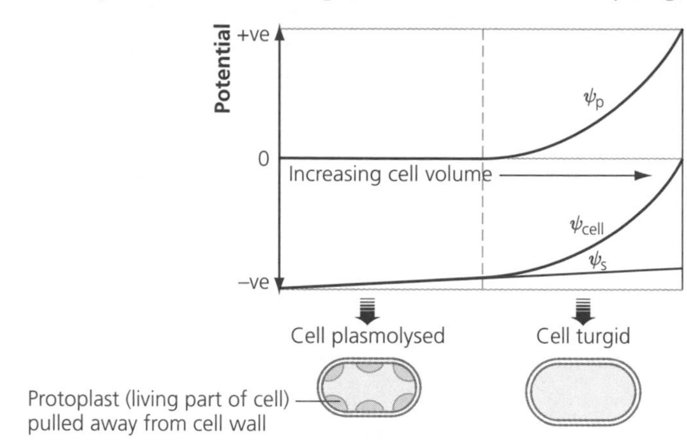

Figure 45 Changes in pressure potential (Ψ_p), solute potential (Ψ_s) and water potential (Ψ_{cell}) of a plant cell as it takes up water osmotically

Active transport across the cell-surface membrane

Active transport causes substances to move across a membrane from a low concentration to a high concentration against the concentration gradient. This requires energy, and the **carrier proteins** involved are referred to as **pumps**. Each carrier protein is specific to just one type of ion or molecule. The substance attaches to a site on the protein (they have complementary shapes) and, with the energy from ATP, the protein changes shape and moves the substance through the membrane.

The movement of substances across the cell-surface membrane is illustrated in Figure 46.

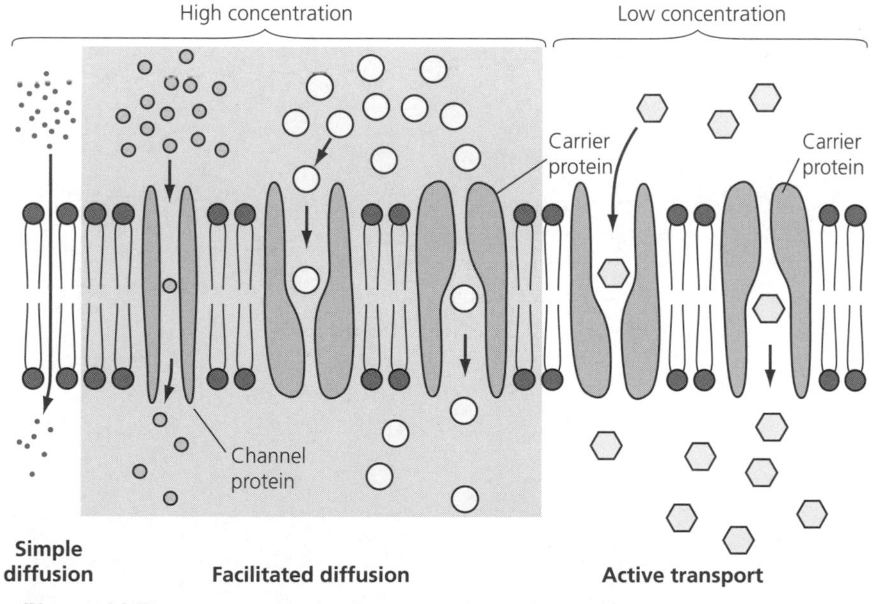

Figure 46 The movement of substances through the cell-surface membrane

Exam tip

Carrier proteins (in facilitated diffusion or in active transport) have a site that is complementary to, and so is specific to, the substance carried. However, *never* refer to this as the active site — that refers to the binding site on enzymes.

Knowledge check 46

State one similarity and one difference between facilitated diffusion and active transport.

Knowledge check 47

Why might a cell benefit from microvilli if carrying out active transport?

Exam tip

Some questions ask you to identify the type of membrane transport illustrated in a diagram, table or graph. Look to see if movement is with or against the concentration gradient and whether ATP is being used.

Cytosis: bulk transport into and out of the cell

Substances can move into and out of a cell without having to pass through the cell-surface membrane. This involves the bulk transport into the cell (**endocytosis**) or out of the cell (**exocytosis**) of substances too large to be transported by protein carriers.

Endocytosis

During endocytosis the cell-surface membrane invaginates and the membrane folds round the substance to form a vacuole or vesicle that enters the cytoplasm while the cell-surface membrane reforms. There are two main types:

- **Phagocytosis** is the uptake of solid particles into the cell within vacuoles. Examples include the ingestion of bacteria by polymorphs (a type of white blood cell) and the removal of old red blood cells from circulation by the Kupffer cells of the liver.
- **Pinocytosis** is the uptake of solutes and large molecules (such as proteins) into the cell within vesicles.

Exocytosis

During exocytosis, **secretory vesicles** move towards and fuse with the cell-surface membrane, releasing their protein contents out of the cell. Exocytosis is also involved in removal of the waste products of digestion from cells.

Cytosis is illustrated in Figure 47.

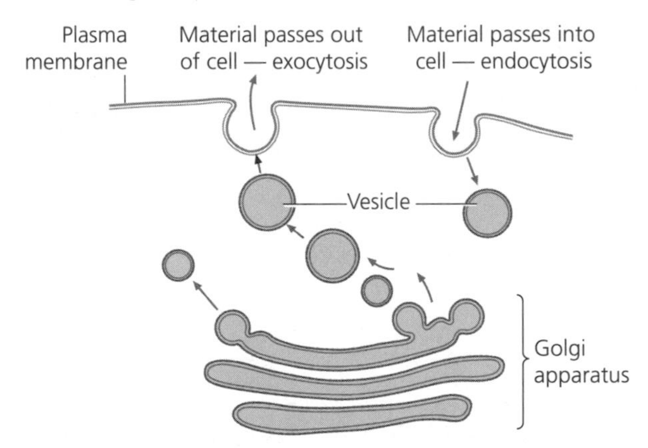

Figure 47 Cytosis: bulk transport

Knowledge check 48

What term would describe the bulk movement of materials in phagocytes?

Practical work

Measure the average water potential of cells in a plant tissue:

- Use a weighing method for potato or other suitable tissue.
- Calculate the percentage change in mass.
- Determine the average water potential from a graph of percentage change in mass against solute potential of immersing solution.

Measure the average solute potential of cells at incipient plasmolysis:

- Use onion epidermis or other suitable tissue.
- Calculate percentage plasmolysis.
- Determine the average solute potential from a graph of percentage plasmolysis against solute potential of the immersing solution — at 50% plasmolysis the average pressure potential is zero.

Exam tip

In your revision, don't forget to include any practical work connected with water potential and solute potential.

Summary

- The basis of membrane structure is the phospholipid bilayer, with hydrophilic heads outermost and hydrophobic fatty acid tails innermost.
- Proteins span the bilayer, or are embedded in or lie on the surface of the bilayer.
- Cholesterol is found in the hydrophobic centre of cell membranes.
- The fluid mosaic model suggests a fluid phospholipid bilayer with scattered protein molecules forming a mosaic pattern.
- Carbohydrates bind to produce glycoproteins and glycolipids on the outer surface of the cell-surface membrane where they form the glycocalyx and are involved in cell recognition and as cell receptors.
- Some proteins may act as enzymes.
- Lipid-soluble (non-polar) molecules and very small molecules (such as O_2 and CO_2) can move by diffusion through the phospholipid bilayer.
- Facilitated diffusion involves the movement of ions through the open pores of channel proteins, and glucose and amino acids via carrier proteins.
- Some water-soluble molecules may also be moved against the concentration gradient by active transport, via carrier proteins that use ATP to pump the molecules across.
- Water moleules can move through the phospholipid bilayer because they are so small, though more often they move through aquaporins.
- Osmosis is the diffusion of water from an area of high water potential to an area of lower water potential through a differentially (or partially) permeable membrane.
- Water potential has two components: solute potential and pressure potential.
- The presence of solutes in solution attracts water molecules and reduces the water potential (makes it more negative) as water is less free to move.
- Plant cells placed in water take in water and develop a pressure potential (turgor pressure) resisted by the cell wall. Animal cells burst when placed in water.
- Exocytosis is the secretion of large molecules (e.g. proteins) as a result of vesicles moving to, and fusing with, the cell-surface membrane.
- Endocytosis allows materials to be taken into the cell, large molecules into vesicles (pinocytosis) or larger materials (e.g. bacteria) into vacuoles (phagocytosis).

The cell cycle, mitosis and meiosis

The formation of new cells, for example in the development of a multicellular organism, involves the production of additional cell contents before a cell can divide. The pattern of events is called the **cell cycle**.

The cell cycle

Actively dividing eukaryotic cells pass through a series of stages known collectively as the cell cycle (Figure 48):

- **interphase** — two **gap (growth) phases** (G_1 and G_2) separated by a **synthesis** (**S**) **phase**
- **mitosis** — a nuclear division during which the chromosomal material is partitioned into daughter nuclei
- **cytokinesis** — the cell divides into two daughter cells

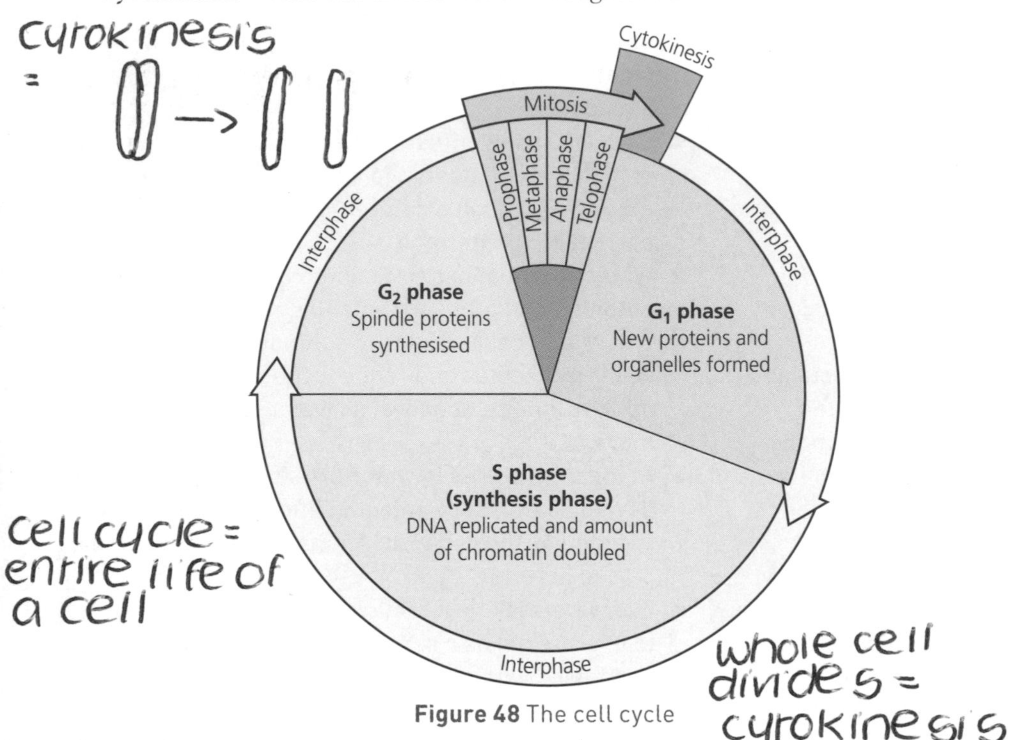

Figure 48 The cell cycle

Interphase

This is an intense period of metabolic activity as the cell synthesises new components such as organelles and membranes, and new proteins and DNA. This takes place in the following phases:

- G_1 phase — the first gap phase. Synthesis of macromolecules, including proteins and nucleotides, occurs and organelles are produced so that the cell increases in size.

Exam tip

The terms cell cycle, mitosis and cell division are quite distinct and should not be confused. Mitosis is one form of nuclear division (the other is meiosis) and is the process by which the nucleus divides. Cell division follows nuclear division and is the process by which the whole cell divides (cytokinesis). The cell cycle represents the entire life of the cell during which it grows, and includes nuclear division and cytokinesis.

- S phase — DNA synthesis occurs. Histones — proteins that bind to and support the DNA within the chromatids — are also produced. The DNA and chromatids formed are identical and remain attached until separated during mitosis (or meiosis).
- G_2 phase — the second gap phase. Proteins such as tubulin are synthesised — tubulin forms the microtubules of the spindle fibres; energy stores are increased; the cell continues to increase in size.

Mitosis

During mitosis different stages are recognised (Figure 49).

Prophase

- The chromatin condenses to form the chromosomes
- The centrioles (in animal cells) move towards opposite poles
- The spindle begins to form
- As each chromosome continues to condense, two chromatids, joined at the centromere, become apparent

Prophase (late)

Metaphase

- The nuclear envelope breaks down
- Spindle fromation is completed as microtubules extend, forming the fibres
- The microtubules of the spindle attach to the centromere of each chromosome
- The chromosomes (chromatid pairs) are moved by the microtubules onto the equator of the spindle

Metaphase

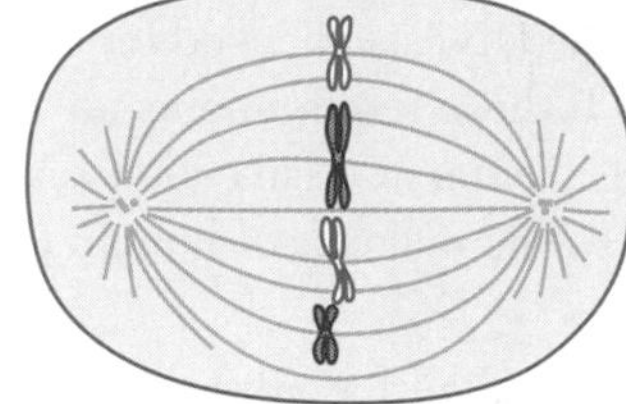

Anaphase

- The centromeres divide
- The spindle fibres pull the centromeres of sister chromatids apart
- The sister chromatids move towards opposite poles

Anaphase

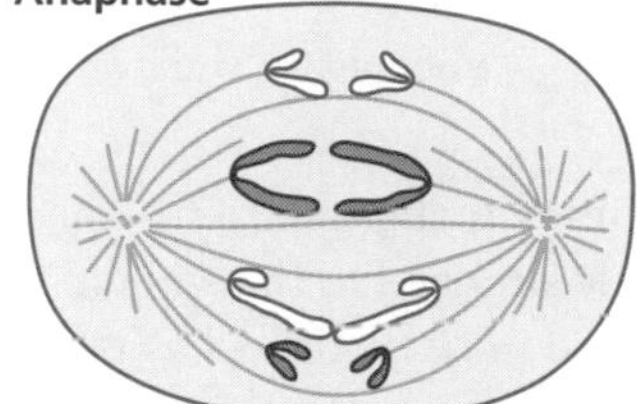

Telophase

- Each chromatid is now a separate chromosome
- The two groups of chromosomes reach opposite poles of the the cell
- A new nuclear envelope forms around each group

Telophase

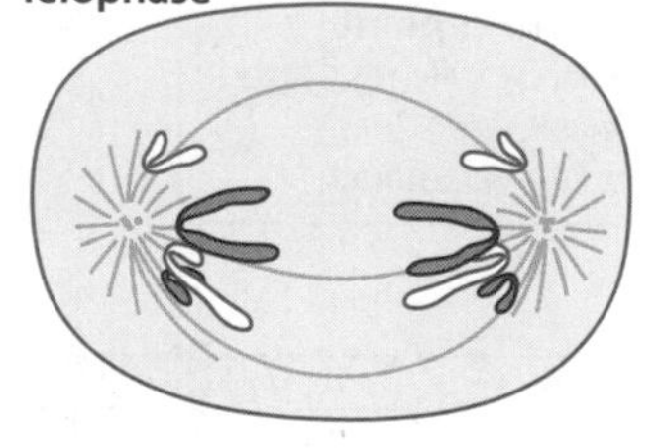

Figure 49 The stages of mitosis

Cytokinesis

At the end of mitosis the cytoplasm is separated and the cell divides during cytokinesis to form two daughter cells. The process differs in animal and plant cells. In the animal cell a **cleavage furrow** forms as protein microfilaments pull the cell

Knowledge check 49

When during the cell cycle does DNA replication take place?

Exam tip

You must be able to distinguish between chromatin and chromosomes. During prophase the chromatin condenses, pulling the DNA into more compact chromosomes (and so preventing breakage). Chromosomes are only visible during nuclear division.

Exam tip

You must be able to distinguish between chromatid and chromosome. Once the centromere has split and the chromatids are separate, they are no longer chromatids but chromosomes, because each now has its own centromere.

surface membrane in along the equator; the furrow deepens and when the membranes fuse the cell is cleaved into two. In plant cells the cell wall prevents cleavage. In the plant cell the Golgi bodies (known as dictyosomes) produce vesicles that collect and fuse together to form an equatorial **cell plate**. The vesicles secrete the material of the middle lamella on each side of which a new cellulose cell wall is laid down.

Knowledge check 50

Describe one difference in nuclear division and one difference in cytokinesis between animal cells and plant cells.

Checkpoints in the cell cycle

It is essential that a cell only divides when:

- sufficient macromolecules and organelles have been assembled so that the cell has grown to an appropriate size
- the DNA is error-free (or is repaired) and accurately replicated
- the chromosomes have been positioned correctly during mitosis

This allows the parent cell to distribute identical DNA and a comparable amount of cellular content to the daughter cells. To ensure this, assessments are made at particular points within the cell cycle. There are three **checkpoints** that must be passed before the cell progresses to the next phase (Figure 50):

1. Towards the end of G_1 the supply of nutrients and growth factors (promoting the synthesis of proteins and other macromolecules) is assessed so that the cell has grown to an appropriate size. The DNA is checked for damage and if damage is detected the DNA is repaired. If these requirements are satisfied then the cell proceeds to the S phase. If not, it enters G_0 (so called 'resting state') where it can remain for days, weeks or years. The majority of human cells are in G_0 and so are not actively dividing.
2. Towards the end of G_2 a check is made that all the DNA has replicated — if it has not, the cell cycle is stopped. Another check is made for DNA damage that may have occurred during replication — once again the cycle may be delayed to repair the DNA.
3. The final checkpoint occurs during metaphase. This check establishes whether the chromosomes have correctly attached to the spindle fibres before anaphase proceeds. Mitosis cannot proceed until this check is passed.

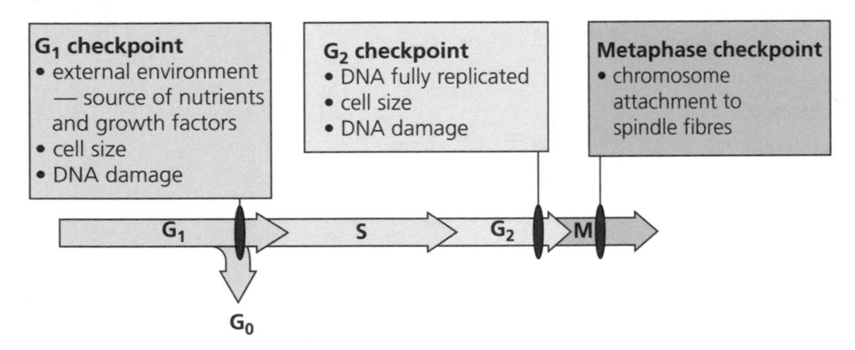

Figure 50 Checkpoints in the cell cycle

Cancer

The passage of a cell-cycle checkpoint is brought about by a series of proteins. If a mutation of the gene that encodes for any of these checkpoint proteins occurs in a cell, then the control mechanisms break down and the cell undergoes repeated, uncontrolled division. A large mass of these cells is called a **tumour**. Tumours can be benign, meaning that they stop growing, are usually encapsulated in a fibrous sheath and do not travel to other locations in the body. If a tumour continues to grow unchecked and uncontrolled, it is termed malignant. A malignant tumour is the basis of a **cancer**. Cells from the primary tumour are shed and carried around the body by the circulatory system, producing secondary tumours (metastases).

Cancer treatments

A number of cancer treatments utilise drugs that target processes within the cell cycle. Different processes may be targeted by **anti-cancer drugs**:

- DNA unzipping is inhibited by the drugs adriamycin and cytoxan. If the two strands of DNA cannot be exposed then DNA replication cannot take place.
- The synthesis of nucleotides (specifically thymine-containing nucleotides) is inhibited by 5-fluorouracil and methotrexate (drugs called antimetabolites).
- Formation of the mitotic spindle is inhibited by taxol and vincristine. Without a spindle, the chromatids cannot separate.

Exam tip

You do not need to learn the names of anti-cancer drugs, although two (5-fluorouracil and vincristine) are noted in the specification. However, be prepared for anti-cancer drugs to be named in exam questions — you are being expected to apply your understanding of the cell cycle.

Knowledge check 51

Which phase in the cell cycle is prevented by 5-fluorouracil? Explain your answer.

Genes, chromosomes and ploidy

Each chromosome contains a long strand of DNA. Specific lengths of DNA along the strand represent the **genes** (which code for the synthesis of polypeptides/proteins), and their positions are called **genetic loci** (singular: locus).

In most plants and animals the cells of the body each contain two sets of chromosomes, which exist in **homologous pairs**. Each member of a pair is similar in size and shape to the other. More importantly, they have the same genetic loci — they possess **alleles** of the same genes (one from each parent). If the alleles on the homologous chromosomes are the same, then the individual is **homozygous** for that characteristic; if they are different, then the individual is **heterozygous**. Cells containing homologous pairs of chromosomes are said to be **diploid** (represented by $2n$).

Diploidy Situation in which a cell has two sets of chromosomes, i.e. a pair of each type.

Chromatids are genetically identical, since DNA replication (during the S phase) produces identical copies. Homologous chromosomes are genetically different, since at least some of the hundreds of genetic loci will possess different alleles.

The relationship between genes, alleles, chromatids and homologous chromosomes is shown in Figure 51.

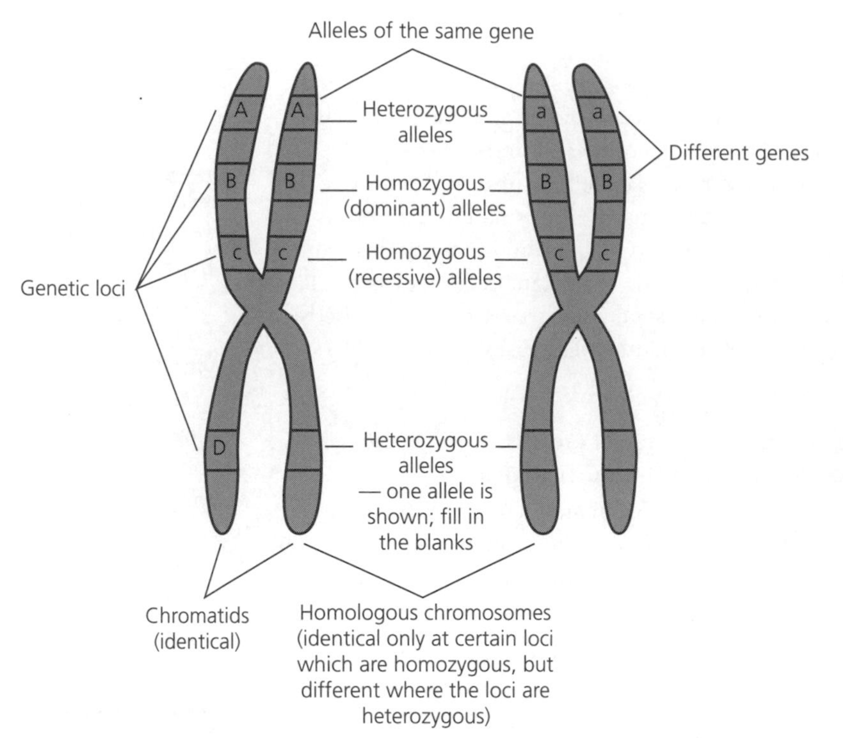

Figure 51 Genes, alleles, chromatids and homologous chromosomes

Knowledge check 52

Complete the alleles present at the **D/d** locus in Figure 51.

Haploidy Situation in which a cell has a single set of unpaired (non-homologous) chromosomes.

A cell that contains only one of each type of chromosome is **haploid** (represented by n). In mammals the gametes are haploid, and diploidy is restored at fertilisation.

Chromosome number and ploidy can be studied by taking a photomicrograph of a cell during nuclear division (at metaphase). This is called a **karyotype**. Arranging the chromosomes into homologous pairs, recognised by their length and centromere position, produces a **karyogram**. A human karyogram contains 23 pairs including a pair of sex chromosomes (Figure 52).

Knowledge check 53

If a cell has 15 chromosomes, is it likely to be diploid or haploid? Explain your answer.

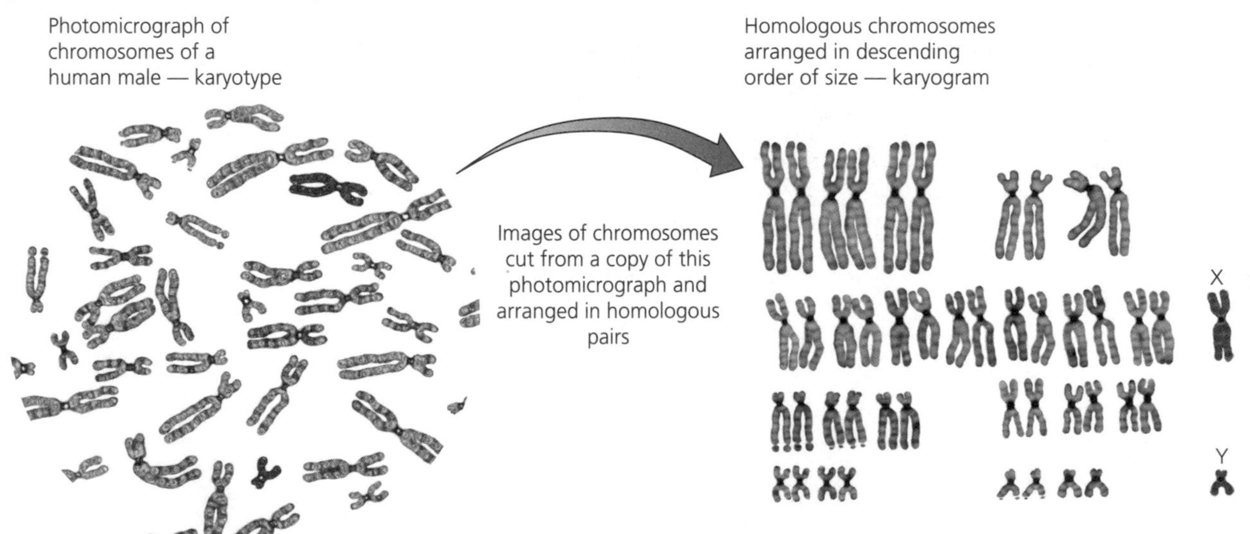

Figure 52 Karyotype and karyogram of a human male

Meiosis

Meiosis occurs only in diploid cells. It involves the separation of homologous chromosomes during a first meiotic division (**meiosis I**) and the separation of chromatids during a second meiotic division (**meiosis II**). The apparatus for these divisions is the same as in mitosis, so the emphasis in Figure 53 is on points specific to meiosis.

Prophase I

- As chromosomes condense it becomes apparent that homologous chromosomes have paired and lie alongside each other; each pair is known as a bivalent
- The chromatids appear: the chromatids in a bivalent are entwined at points called chiasmata (singular: chiasma)
- The chromatids may break at chiasmata and rejoin with a different chromatid, resulting in crossing over or recombination

Prophase I (late)

Metaphase I

- The bivalents move to the equator of the spindle
- Each chromosome of the pair becomes attached to a spindle fibre by its centromere

Metaphase I

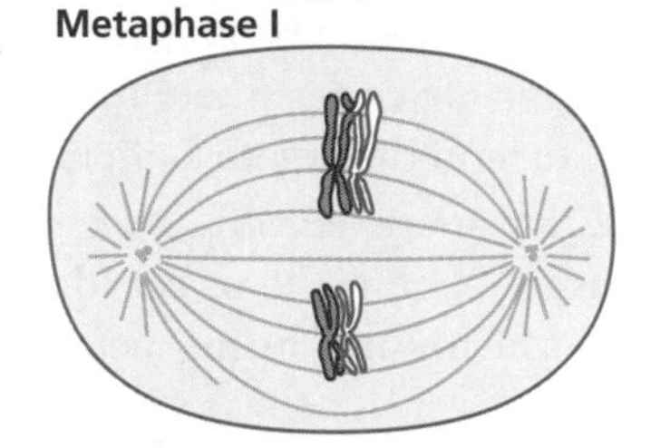

Anaphase I

- Pulling by the spindle fibres causes the homologous chromosomes to move apart towards opposite poles
- The homologous chromosomes are separated; each chromosome still consists of two chromatids

Anaphase I

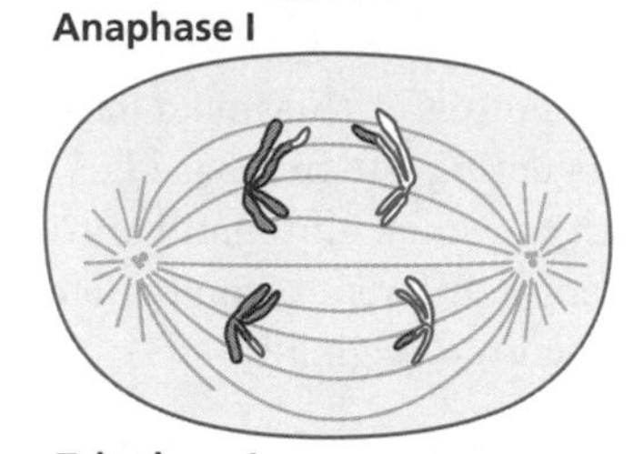

Telophase I

- Chromosomes reach opposite poles of the cell.
- A nuclear membrane forms around each separate group of chromosomes; each nucleus contains the haploid number of chromosomes

Telophase I

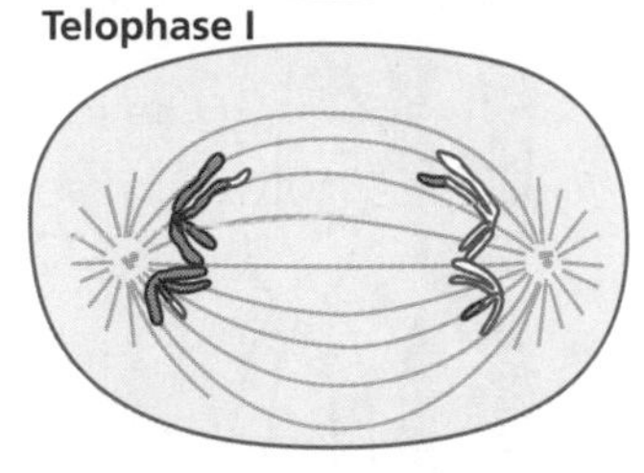

Figure 53 The stages of meiosis I

Cytokinesis after meiosis I produces two daughter cells. Within each, meiosis II follows:

1. New spindles begin to form at right angles to the old spindle (**prophase II**).
2. Chromosomes consisting of pairs of chromatids (now different because of crossing over) are arranged along the equator (**metaphase II**).
3. Sister chromatids are split at the centromere and pulled to opposite poles (**anaphase II**).
4. Each group of separated chromosomes becomes enclosed within a nuclear envelope (**telophase II**).

Exam tip

Be aware that meiosis I is about the separation of the homologous chromosomes; meiosis II is about separation of the chromatids. So the chromosome number is halved after meiosis I, and the cells formed at this stage are haploid cells.

Knowledge check 54

In prophase I of meiosis how many

a chromatids

b chromosomes

are there in one bivalent?

Exam tip

Use the appearance and number of chromosomes at *anaphase* to identify whether a division is mitotic or meiotic. If the chromosomes are double structures (two chromatids) it can only be anaphase I of meiosis. If they are single structures, it could be mitosis or anaphase II of meiosis. In that case look at how many there are. If the diploid number is moving to each pole, it must be mitosis; if the haploid number is moving to each pole, it must be meiosis II.

Cytokinesis follows. The overall result of meiosis is the production of four haploid daughter cells, each of which is genetically different from the others.

The significance of mitosis

Mitosis produces genetic constancy:

- The daughter cells possess the *same* chromosome number as each other and as the parent cell. Mitosis can occur in either diploid or haploid cells.
- The daughter cells are *genetically identical*. Mitosis has a key role in asexual reproduction, producing genetically identical individuals (**clones**).

Exam tip

When you are talking about the products of mitosis, you should always say 'genetically identical cells', not just 'identical cells'.

The significance of meiosis

Meiosis produces change:

- Meiosis is the type of nuclear division that transforms the diploid condition into the haploid condition. This is vital in life cycles where **fertilisation** involves the fusion of gametes (haploid cells) to form the zygote (a diploid cell).
- Meiosis produces daughter cells that are *genetically different*. This genetic variation occurs as a result of **crossing over** of chromatid pieces (during prophase I) and of the **independent assortment of bivalents** (during metaphase I).

Crossing over

Crossing over (Figure 54) occurs as a result of chiasmata formation between the chromatids of the homologous pairs during late prophase I. A piece of chromatid from one chromosome swaps places with a piece of chromatid of the homologous partner. It results in each chromosome having a different combination of alleles (called **recombinants**) from that which occurred originally.

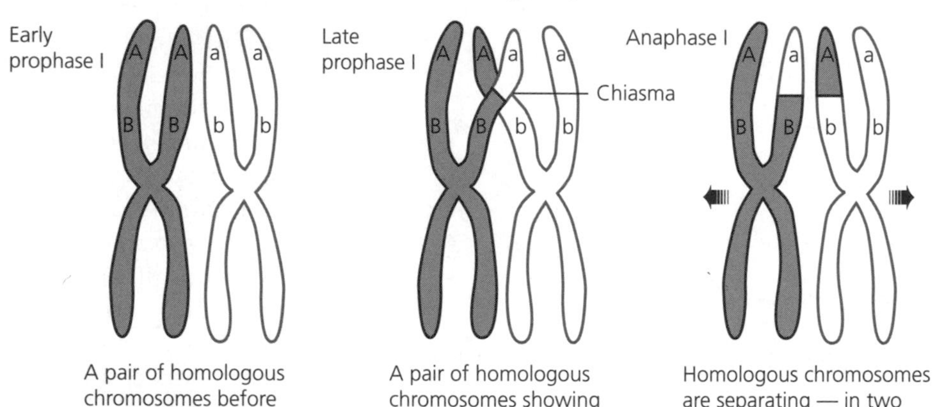

Figure 54 Crossing over and genetic variation

Knowledge check 55

After a single cross-over how many chromatids in a bivalent have a new combination of alleles?

Independent assortment

During metaphase I bivalents are arranged at *random* on the equator of the spindle. This means that the orientation of any one homologous pair is not dependent on the orientation of any other pair. When the homologous chromosomes are pulled apart at anaphase I, a chromosome of one pair is equally likely to be separated along with either member of any other homologous pair.

Independent assortment is illustrated in Figure 55.

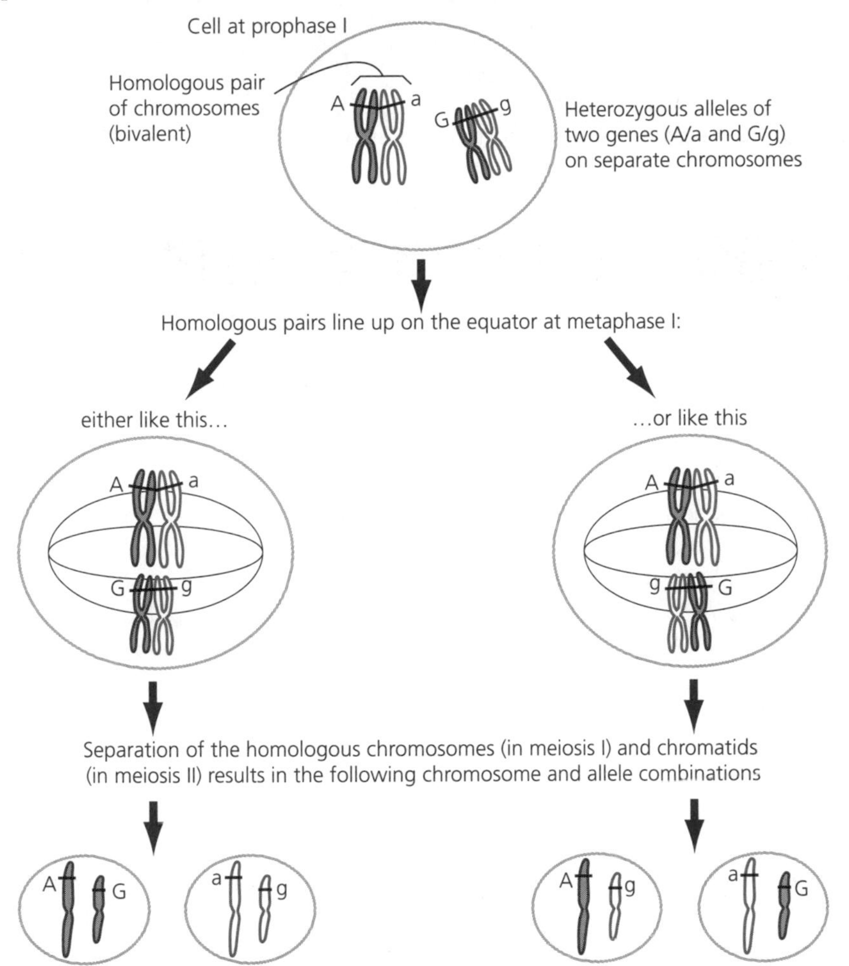

Figure 55 Independent assortment and genetic variation

Mitosis and meiosis are compared in Table 15.

Table 15 Comparisons between mitosis and meiosis

Mitosis	Meiosis
One division, producing two daughter cells	Two divisions, producing four daughter cells
Parent cell may be either diploid or haploid; daughter cells have the same chromosome number as the parent cell	Parent cell is always diploid; daughter cells are haploid
Homologous chromosomes (if the parent cell is diploid) do not associate during prophase	Homologous chromosomes pair, forming bivalents during prophase I
No chiasmata formation	Chiasmata form between the chromatids of the homologous chromosomes during prophase I
Daughter cells are genetically identical	Daughter cells are genetically different

Knowledge check 56

If diploid cells ($2n = 6$) divide by meiosis, how many different combinations of chromosomes would it be possible to find among the haploid cells produced?

Exam tip

You will be penalised if certain words are not spelled correctly. For example, if you write 'meitosis' or 'miosis' you will not get a mark because the examiner won't know which process you are referring to.

Practical work

Prepare and stain root tip squashes, and examine prepared slides or photographs of the process of mitosis:

- Recognise chromosomes at different stages of cell division.
- Identify the stages of mitosis.

Examine prepared slides or photographs of the process of meiosis:

- Identify the structures visible during meiosis and the different stages of meiosis.

Summary

- Eukaryotic cells exhibit a cell cycle consisting of interphase, nuclear division (mitosis or meiosis) and cytokinesis (cell division).
- Interphase is sub-divided into: a gap phase (G_1) of biosynthesis and increase in organelle numbers; a DNA replication phase (S); and a second gap phase (G_2), when proteins necessary for nuclear division are produced.
- Mitosis produces two daughter nuclei, with the same number of chromosomes, and which are genetically identical.
- Mitosis involves four phases: prophase (chromosomes condense and spindle assembly commences); metaphase (chromosomes are assembled on the equator of the spindle); anaphase (chromatids are separated); and telophase (new nuclei form).
- Mitosis is the type of nuclear division that occurs during growth and body repair.
- Nuclear division is followed by cytokinesis, which involves:
  - in animal cells, the plasma membrane forming a constriction that eventually 'pinches off' the cytoplasm, forming two new cells
  - in plant cells, a cell plate forming in the centre of the cell, which grows outwards and forms two new cell walls that separate the daughter cells
- Cell cycle checkpoints, at the end of G_1 and G_2 and during metaphase, ensure that the DNA has been accurately duplicated and the cell contents evenly divided.
- Defects in checkpoints may lead to uncontrolled cell division and the formation of a cancer.
- Anti-cancer drugs halt cell division by inhibiting a vital process during the cell cycle.
- Meiosis is the type of nuclear division in a diploid cell that produces four haploid cells, each of which will vary genetically.
- In mammals, meiosis produces the gametes.
- Meiosis involves two nuclear divisions: in meiosis I homologous chromosomes are separated, while in meiosis II chromatids are separated. Each division has the phases prophase, metaphase, anaphase and telophase.
- Meiosis produces genetically variable haploid cells through the processes of crossing over and independent assortment.

Tissues and organs

Animals and plants are multicellular — they are made up of large numbers of cells. Cells become specialised according to their function. **Tissues** are groups of cells of the same type performing a particular function. For example, in plants mesophyll is the photosynthetic tissue in leaves; in animals muscle is a contractile tissue allowing movement. **Organs** are structures made of several tissues that work together to carry out a number of functions. The **leaf** is an organ of photosynthesis with not just photosynthetic tissue, but other tissues that provide for gaseous exchange, transport and protection. The **ileum** is the organ, in the small intestine, that is concerned with the final stages of digestion, the absorption of the products of digestion, and the movement of undigested material along to the large intestine. The ileum works with other organs within the digestive system — an example of an **organ system**.

The ileum

The ileum is the region of the small intestine where digestion is completed and where most absorption of the products of digestion occurs. There is a vast surface area for digestion and absorption provided by:

- folds in the inner surface of the intestinal wall
- projections called **villi** (singular: villus) that are present on the folded surface of the wall
- microscopic projections called **microvilli** on the cell-surface membranes of **columnar epithelial cells** that line the villi

Structurally the ileum consists of tissues in distinct layers: **mucosa**; **muscularis mucosa**; **submucosa**; **muscularis externa** and **serosa** (outermost) (Figure 56).

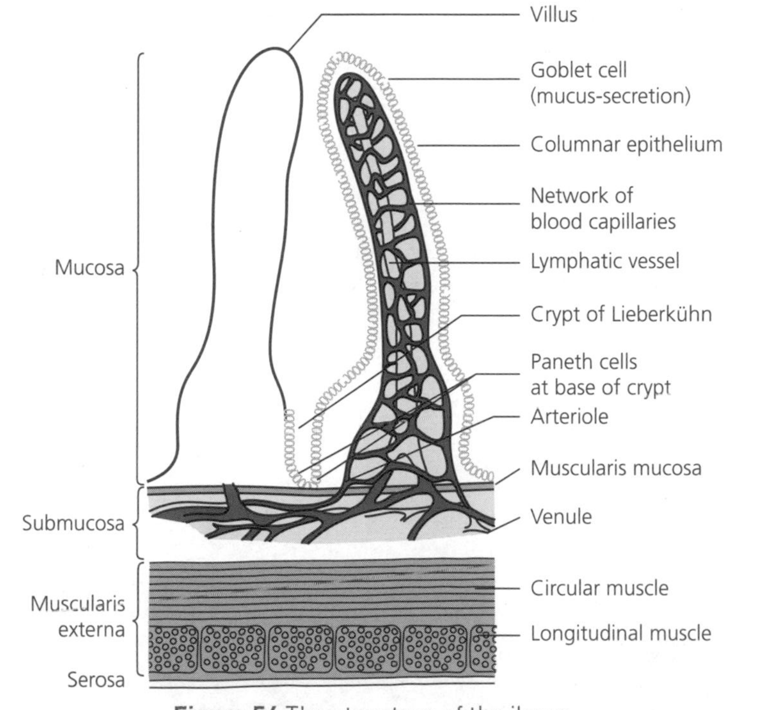

Figure 56 The structure of the ileum

The functions of the tissues are shown in Table 16.

Table 16 The functions of tissues in the ileum

Tissue	Function
Columnar epithelium (within the mucosa)	This layer has column-shaped cells and lines the intestine. On their free surfaces the cells have microvilli, forming a brush border. Since digestive enzymes are bound to the membrane of the microvilli, this provides a huge surface area for digestion and for the absorption of the products of digestion. Some substances are taken up partly by diffusion and partly by active transport; others are taken up by pinocytosis. There are numerous mitochondria to aid active transport. The cells of the epithelium are short-lived (see crypts of Lieberkühn).
Goblet cells (within the epithelium)	These cells secrete mucus. Mucus is slimy. It protects the epithelium from the action of digestive enzymes and lubricates the lining as solid material is pushed along.

Exam tip

Students often confuse villi with microvilli. Villi are 1-mm projections of the wall of the mucosa and each contains thousands of cells. Microvilli are 0.6-μm projections of the cell-surface membrane of the epithelial cells lining the mucosa.

Knowledge check 57

Explain why mucus is needed to protect the cells lining the ileum from protein-digesting enzymes.

Knowledge check 58

State one way in which the ileum is adapted to churn the food.

Exam tip

Make sure that you distinguish between absorption (taking soluble molecules into the body) and assimilation (incorporating absorbed molecules into body tissues).

→

Tissue	Function
Villi (within the mucosa)	These finger-like projections increase the surface area for the absorption of the products of digestion. The villi contain blood capillaries into which amino acids and monosaccharides are absorbed, and lacteals (blind-ending lymph vessels) into which fats are absorbed.
Crypts of Lieberkühn (within the mucosa)	These intestinal glands are found at the bases of the villi. The cells along the sides secrete mucus. The cells (stem cells) lining the bottom of the crypts are in a state of continuous division; new cells are continuously being pushed up by the division of cells deeper down. After a life of several days within the epithelium, the cells are pushed to the tips of the villi where they are sloughed off. Paneth cells are also present at the base of the crypts. Their function is to defend the actively dividing cells against microbes in the small intestine.
Muscularis mucosa	The muscle fibres contract to cause movement of the villi, so improving contact with the products of digestion.
Submucosa	The submucosa contains blood vessels, including venules of the hepatic portal vein (carrying blood to the liver) and lymphatic vessels, supported by connective tissue.
Muscularis externa	The muscularis externa consists of circular muscle (innermost) and longitudinal muscle. Contractions of longitudinal muscle causes pendular movement of the gut while contraction of circular muscle may result in local constrictions, both of which churn the food. Coordinated contractions of the circular muscle push food along the gut by peristalsis.
Serosa	This outer layer of connective tissue serves to protect and support the gut.

The leaf

The leaf has a large surface area, which maximises the absorption of light for photosynthesis. It is thin, so photosynthesising cells are not far from the leaf surfaces where light absorption and gaseous exchange occur.

Structurally, the leaf consists of epidermal layers either side of a middle layer of mesophyll and vascular tissues (Figure 57).

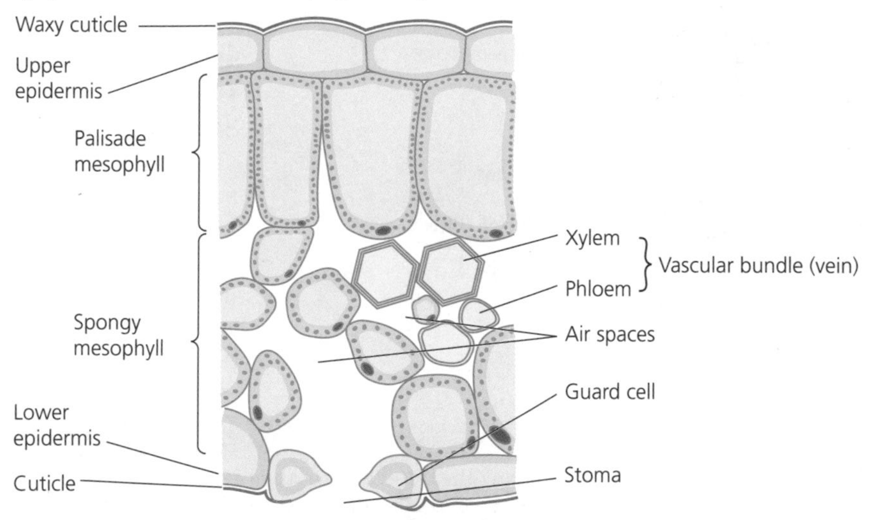

Figure 57 The structure of the leaf

Exam tip

Remember that plant cells respire all the time, but only plant cells with chloroplasts photosynthesise and then only when light is available.

Knowledge check 59

Explain how the palisade mesophyll is adapted to carry out its function.

Knowledge check 60

What is the advantage to a plant of being able to control the opening and closing of stomata?

The functions of the tissues are shown in Table 17.

Table 17 The functions of tissues in the leaf

Tissue	Function
Upper epidermis	The cells of the upper epidermis lack chloroplasts since their role is protective. They secrete a waxy cuticle that provides waterproofing and reduces water loss.
Palisade mesophyll	The palisade layer, in the upper half of the leaf, has layers of tightly packed cells, each with abundant chloroplasts. It is adapted for maximal light absorption. This is the main photosynthetic region of the leaf.
Spongy mesophyll	The mesophyll in the lower half of the leaf contains large air spaces. Gaseous exchange between these air spaces and the atmosphere can take place via numerous pores (stomata). Spongy mesophyll cells also contain chloroplasts and are photosynthetic.
Xylem vessels	Xylem vessels supply the leaf with water and inorganic ions.
Phloem sieve tubes	Phloem sieve tubes transport sucrose away from the leaf.
Lower epidermis	The cells lack chloroplasts. The cuticle secreted on the lower surface is thinner than that on the upper surface since it is not exposed directly to the Sun.
Stomata	The lower epidermis contains numerous stomata, which allow gaseous exchange. They also allow water vapour to diffuse easily out of the leaf. Each stoma (singular of stomata) is surrounded by a pair of guard cells, which cause it to close at night and so water loss by transpiration is minimised.

Skills development

Drawing skills

In an AS paper you may be asked to make a labelled drawing from a photograph. A **drawing skills question** could ask you to 'draw a block diagram to show the tissue layers shown in the photograph'. You will be tested on:

- your ability to identify the tissue layers and construct a block drawing showing all the tissues obvious in the photograph
- how true the drawing is to the photograph provided (and not just a textbook diagram of the feature) and the drawing being the same magnitude as the photograph (or having scale added if appropriate)
- the position and proportionality of the tissue layers
- the quality of the drawing so that clear, smooth and continuous lines are drawn

As an example, Figure 58 is a labelled drawing of a photograph of the midrib region of a privet leaf that shows the tissue layers.

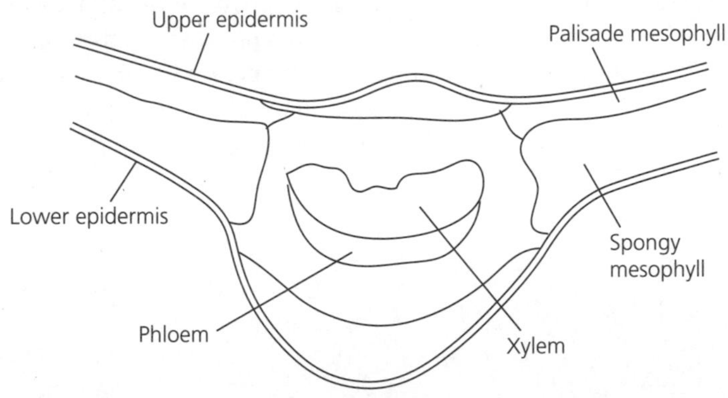

Figure 58 A block diagram of the midrib region of a privet leaf

Practical work

Examine stained sections of the ileum using the light microscope and electron micrographs or photomicrographs:

- Recognise the villi (and associated blood capillaries and lacteals), crypts of Lieberkühn (and Paneth cells), mucosa, columnar epithelium, goblet cells, muscularis mucosa, submucosa, muscularis externa and serosa.

Examine sections of a mesophytic leaf using the light microscope or photomicrographs:

- Recognise the epidermal layers, waxy cuticles, palisade mesophyll, chloroplasts, spongy mesophyll, vascular vessels with xylem and phloem, and guard cells and stomata.

Make accurate drawings of sections of the ileum and the leaf to show the tissue layers:

- Draw block diagrams of tissues in the ileum and the leaf.

Exam tip

If you are asked to explain how the structure of a tissue or organ is adapted to carry out its function, you need to state the feature and explain how it enables the tissue or organ to do its job.

Summary

- An organ is a structure that consists of several different tissues, each performing different functions that contribute to the overall functioning of the organ.
- The ileum is the organ in mammals in which the final stages of digestion and most absorption take place.
- The absorptive surface is the columnar epithelium, with microvilli, on the villi of the mucosa layer.
- Monosaccharides and amino acids are absorbed into blood vessels but the products of fat digestion enter the lacteals (lymphatic vessels).
- Crypts of Lieberkühn possess stem cells, which produce new epithelial cells, and Paneth cells, which have an antimicrobial function.
- Contraction of the muscularis mucosa imparts a wafting action to the villi, ensuring that they are in contact with freshly digested food.
- The submucosa contains blood and lymphatic vessels within connective tissue.
- The muscularis externa consists of circular and longitudinal muscles, which help to churn the food and move it along by peristalsis.
- The serosa covers the outside of the intestine and has a protective function.
- The leaf is the organ of photosynthesis in higher plants.
- The upper and lower epidermis are colourless and protect the leaf from damage and infection and, with the secretion of a waxy cuticle, from dehydration.
- Stomata are found mostly in the lower epidermis and open during the day to allow the diffusion of CO_2 into the leaf.
- The palisade mesophyll consists of columnar cells containing many chloroplasts and represents the main photosynthetic layer.
- The spongy mesophyll contains large air spaces, which allow diffusion of gases through the leaf and contain chloroplasts for photosynthesis.
- The veins in the leaf contain the transport tissues xylem and phloem.

Questions & Answers

The examination

The AS Unit 1 examination constitutes 37.5% of the AS award; and, since AS represents 40% of A-level, contributes 15% to the final A-level outcome. The paper lasts 1 hour 30 minutes and is worth 75 marks. In Section A (60 marks) all the questions are structured, though with a variety of styles. In Section B (15 marks) there is a single question, possibly with several parts, to be answered in continuous prose.

Examiners construct papers to test different assessment objectives (AOs). In the AS Unit 1 paper the approximate marks allocated for each AO are:

- AO1 Knowledge and understanding 27 marks
- AO2 Application of knowledge and understanding 31 marks
- AO3 Analysis, interpretation and evaluation of scientific information, ideas and evidence 17 marks

Skills assessed in questions

The questions in this section test the different assessment objectives. While some assess straightforward knowledge and understanding (AO1), don't be surprised to find something novel — you are being asked to *apply* your understanding (AO2). Some questions will ask you to evaluate experimental and investigative work (AO3).

Since mathematical skills are an important element of biology, the questions include a variety of calculations and graphical work.

Quality of written communication, including accurate use of scientific terms, is assessed in Section B.

About this section

This section consists of questions covering the range of topics covered in the Content Guidance. Following each question there are answers provided by two students of differing ability. Student A consistently performs at grade A/B standard, allowing you to see what high-grade answers look like. Student B makes a lot of mistakes — ones that examiners often encounter — and grades vary between C/D and E/U.

Each question is followed by a brief analysis of what to look out for when answering the question (shown by the icon ⓔ). All student responses are then followed by comments (preceded by the icon ⓔ). They provide the correct answers and indicate where difficulties for the student occurred, including lack of detail, lack of clarity, misconceptions, irrelevance, poor reading of questions and mistaken meanings of examination terms. The comments suggest areas for improvement.

In using this section try the questions before looking at the students' responses or the comments, which you can then use to mark your work. Check where your own answers might have been improved.

The CCEA biology specification is available from www.ccea.org.uk. Apart from the *subject content*, you must familiarise yourself with the *mathematical skills* shown in section 4.7 and the *command terms* used in examinations (such as *explain, describe* and *suggest*) shown in Appendix 1. The website will also allow you to access *past papers* and *mark schemes*.

Section A Structured questions

Biological molecules

Question 1 Carbohydrates

(a) **A key was constructed to allow the identification of the following carbohydrates: cellulose, fructose, glucose, starch, sucrose. Use the key to identify each of the carbohydrates A to E.** (3 marks)

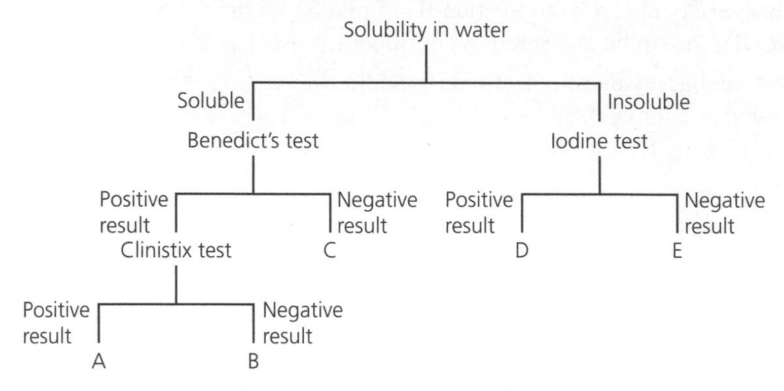

(b) **Name the group of sugars to which fructose and glucose belong.** (1 mark)

(c) **Name the sugars found in nucleic acids and describe how they are structurally different from glucose and fructose.** (2 marks)

(d) **Sugars may be added to soft drinks as sweeteners, though the level of sweetness provided will depend on the sugar used. Glucose has only 75% of the sweetness of sucrose but fructose is twice as sweet as glucose. Determine how much fructose must be added to replace 24 g of sucrose. Show your working.** (2 marks)

Total: 8 marks

e This question is quite varied and includes a part on practical work and a part testing mathematical skills. In part (a) you are presented with a problem-solving situation involving the identification of five carbohydrates — study the key carefully since you will not have seen it before. You are being asked to apply your understanding of biochemical tests, though there are only 3 marks available because Benedict's and the iodine tests should be very familiar to you. In parts (b) and (c) you are being tested on your knowledge of monosaccharides — you need to give precise and full answers. In part (d) you are given a numeric problem to solve. If the solution does not come readily, you should return to this part later when your confidence has grown.

Student A

(a) **A** glucose
B fructose } ✓
C sucrose

D starch ✓
E cellulose ✓ a

(b) They are hexose sugars. ✓ b

(c) Ribose and deoxyribose. ✓ They are not used in respiration. ✗ c

(d) sweetness of fructose = 150% (75% × 2) sweetness of sucrose ✓

24 g ÷ 150% = 16 g ✓ d

e 7/8 marks awarded a All correct, for 3 marks. (Four correct responses would have been awarded 2 marks while three were required for a single mark because the question involves understanding gained prior to A-level.) b Correct, for 1 mark. c The answer includes both sugars present in nucleic acids but unfortunately does not answer the second part of the question because 'not used in respiration' is not a structural difference. Questions need to be read carefully. d Student A has worked out the relationships well and seems at ease with numerical problems. Both marks are awarded.

Student B

(a) A fructose ✗
B glucose ✗
C sucrose
D starch } ✓
E cellulose a

(b) monosaccharides ✗ b

(c) deoxyribose ✗

Deoxyribose is a 5-carbon sugar, while glucose and fructose are 6-carbon sugars ✓ c

(d) 75% of 24 g = 16 g ✗

16 ÷ 2 = 8 g ✓ d

e 3/8 marks awarded a Student B understands that sucrose is a non-reducing sugar and that iodine tests for starch. However, he/she cannot distinguish fructose and glucose and, in particular, that Clinistix is a specific test for glucose. 1 mark scored. b This is correct but not precise enough. The correct answer is given in part (c) but should be presented here for the mark to be gained. c Deoxyribose is the sugar in DNA but RNA — the other nucleic acid — contains ribose, which should have been included to give a full answer. Nevertheless, the distinction from the 6-C hexose sugars is understood. 1 mark gained. d If glucose has 75% of the sweetness of sucrose, then more of it is required to give the same level of sweetness. So 32 g of glucose equates to 24 g of sucrose, which equates to 16 g (32/2) of fructose, since fructose is twice as sweet as glucose. Student B has not got the first relationship, but understands to divide by 2 for the link between glucose and fructose and so earns 1 mark.

Question 2 Lipids

The diagram below shows the structure of one type of lipid, a triglyceride.

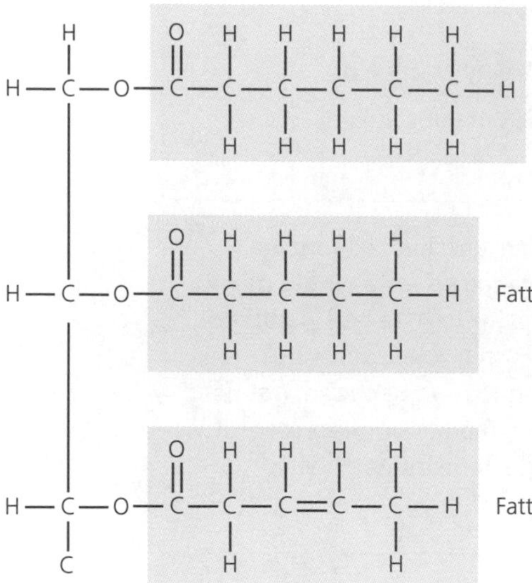

Fatty acid A

Fatty acid B

Fatty acid C

(a) **Name the following**

(i) **the molecule to which the fatty acids are joined** (1 mark)

(ii) **the type of reaction used to attach the fatty acids to this molecule** (1 mark)

(iii) **the bond between the fatty acids and this molecule** (1 mark)

(b) **Use the diagram to explain what is meant by an unsaturated fatty acid.** (1 mark)

(c) **Oils and fats are triglycerides. Describe the difference in structure between oils and fats that results in oils being liquid and fats solid at room temperature.** (2 marks)

(d) **Describe how a phospholipid differs in structure from a triglyceride.** (1 mark)

(e) **Describe the characteristics of phospholipid molecules that determine their arrangement in a cell membrane.** (2 marks)

Total: 9 marks

e This is a fairly straightforward question. It assesses your understanding of a variety of lipids. You need to be careful in wording your answers to ensure that you provide the necessary detail for marks to be awarded.

Student A

(a) **(i)** glycerol ✓

(ii) condensation ✓

(iii) ester ✓ a

(b) Fatty acid C is unsaturated. ✗ b

(c) Oils tend to consist of unsaturated fatty acids with relatively short hydrocarbon chains. ✓ Fats contain saturated fatty acids with long hydrogen chains. ✓ c

(d) A phospholipid consists of glycerol bonded to two fatty acids and a phosphate instead of three fatty acids as in a triglyceride. ✓ **d**

(e) The phosphate group at the glycerol end is charged, making that end hydrophilic, while the hydrocarbon chains are not water soluble and so are hydrophobic. ✓ As a result, phospholipids form bilayers with the hydrocarbon chains innermost and away from water and the phosphate ends in contact with the cytosol on the inside or the extracellular fluid. ✓ **e**

e **8/9 marks awarded** **a** All correct, for 3 marks. **b** This is true but does not answer the question, which asks for an explanation of what an unsaturated fatty acid is (see student B's answer). No mark is awarded. **c** Student A has provided more than expected (reference to the relative lengths of hydrocarbon chains is a bonus) and gains 2 marks. **d** Correct, for 1 mark. **e** A full answer for 2 marks.

Student B

(a) (i) glycerol ✓

(ii) hydrolysis reaction ✗

(iii) glycosidic bond ✗ **a**

(b) It has a double bond between carbon atoms. ✓ **b**

(c) Oils contain unsaturated fatty acids. ✓ **c**

(d) A phospholipid has a phosphate group joined to the glycerol. ✗ **d**

(e) The phosphate makes the glycerol end hydrophilic while the hydrocarbon chains of the fatty acids are hydrophobic. ✓ **e**

e **4/9 marks awarded** **a** Sub-part (i) is correct. In (ii) student B has confused hydrolysis ('breaking down with water' reaction) and condensation ('building up with the release of water' reaction). In (iii) the bond presented is that which joins monosaccharides — perhaps there was a false association between glycosidic and glycerol. You need to be careful when learning the names of biochemicals. **b** This is correct, for 1 mark. **c** This is good, for 1 mark, but the answer is not complete because it neglects to say that fats contain saturated fatty acids. **d** This is not wrong — it is simply not precise enough. In a phospholipid a phosphate replaces a fatty acid. **e** This is not a full enough answer because there is no reference to the orientation of the molecules in the membrane. Only 1 mark is awarded.

Question 3 Proteins

(a) The protein somatostatin consists of a single, short chain of amino acids.

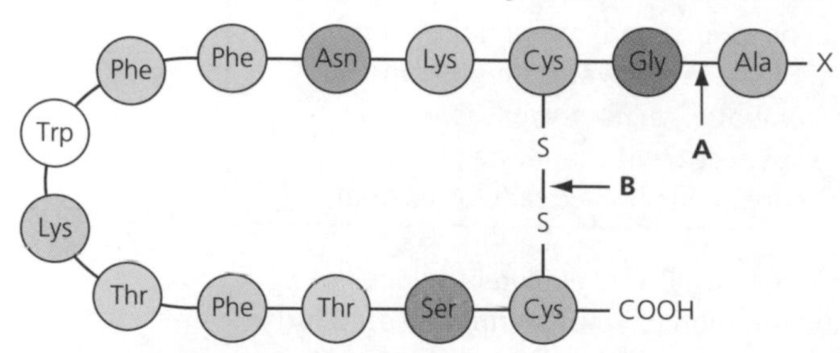

(i) Name the bonds shown at A and B. (2 marks)

(ii) Name the chemical group at X. (1 mark)

(iii) Two of the amino acids shown are alanine (Ala) and glycine (Gly). Explain how they differ. (1 mark)

(b) Explain, fully, the terms quaternary and conjugated in relation to the structure of haemoglobin. (4 marks)

Total: 8 marks

e Make sure that you study any diagrams in the exam paper carefully. Part (a) asks you to identify features in the diagram. The diagram should also act as a stimulus to the next part, which tests your overall understanding of protein structure. Notice that in part (b) you are asked to *fully* explain the terms quaternary and conjugated. You should also note that the answer is worth 4 marks and so a detailed answer is required.

Student A

(a) (i) A — peptide bond ✓; B — disulfide link ✓ a

(ii) NH_2 ✓ a

(iii) They would have different R groups. ✓ a

(b) Haemoglobin is a quaternary protein as it consists of four polypeptides ✓, two α-chains and two β-chains ✓. It is also conjugated, as each polypeptide has a prosthetic group attached ✓; specifically, an iron-containing haem group ✓. a

e **8/8 marks awarded** a All answers correct, for full marks.

Student B

(a) (i) A — condensation bond ✗; B — ? ✗ a

(ii) amino group ✓ b

(iii) They have different structures. ✗ c

(b) Quaternary means the protein consists of more than one polypeptide. ✓ Conjugated means that a non-protein part is involved. ✓ d

e **3/8 marks awarded** **a** The answer for A is not correct, since condensation is a type of reaction and takes place in the formation of polysaccharides and triglycerides as well as polypeptides. Regarding B, it is important that you recall facts like the bonds in the tertiary structure of a protein (such as the disulfide bridge) as they are a source of easy marks. Student B scores no marks. **b** Correct, for 1 mark. **c** This does not have sufficient detail to be awarded the mark. Student B needed to mention that R groups differ between the amino acids. **d** Student B has not provided a detailed answer and has ignored the requirement to explain the terms in relation to the structure of haemoglobin. 2 marks scored.

Question 4 Prions

(a) A normal protein, PrP^{C}, found in the cell-surface membrane, can be induced to alter its form into the disease-causing prion protein, PrP^{Sc}.

(i) Describe how the shapes of the two forms of protein differ. (2 marks)

(ii) Explain why PrP^{Sc} is described as disease causing. (2 marks)

(iii) Name the disease caused by PrP^{Sc} in humans. (1 mark)

(b) A case of BSE was detected on a farm in County Louth in 2015.

(i) The *Irish Times* had the headline 'Three possible causes being explored in BSE investigation'. Suggest what these were. (3 marks)

(ii) A total of 67 cattle, born and raised on the same farm, tested negative for BSE. What does this suggest about the cause of the outbreak? (1 mark)

Total: 9 marks

e Part (a) is a simple matter of knowing facts about prion proteins and prion disease. You will need to be careful with the spelling for (iii). Part (b) requires you to apply your understanding of how prion disease may occur in cattle.

Student A

(a) (i) The secondary structure of PrP^{C} is mainly composed of α-helices ✓, whereas PrP^{Sc} is rich in β-sheets ✓. **a**

(ii) The prion protein PrP^{Sc} has the ability to cause normal PrP^{C} to change its secondary structure to the PrP^{Sc} form. ✓ As the misshapen proteins accumulate, the way in which the proteins interconnect alters. This results in neurone death, particularly in the brain, where holes develop. ✓ **a**

(iii) Creutzfeldt-Jacob disease ✓ **a**

(b) (i) Transmission through consumption of food infected with the prion protein or otherwise from the environment. ✓ Inheritance of a genetic mutation to the allele that codes for the disease form PrP^{Sc}. ✓ Sporadic or spontaneous formation of the PrP^{Sc} form. ✓ **a**

(ii) Since no other animal tested positive for the presence of the prion protein it is very unlikely that the disease was acquired from feed or the farm environment. ✓ **a**

e **9/9 marks awarded** **a** All answers correct, for full marks.

Student B

(a) (i) A normal protein, PrP^C, has a mostly helical primary structure, ✗ while the prion protein PrP^{Sc} has extensive pleated sheets. ✓ a

(ii) Misfolded PrP^{Sc} can cause the normal protein to change its shape to the disease-causing PrP^{Sc} form. ✓ As a result, there is a build-up of abnormal proteins, which result in brain cell death. ✓ b

(iii) Jacob Cruzfelt disease ✗ c

(b) (i) Acquired through eating infected feed; ✓ through prions dormant in the soil entering an open wound; through an injection with a contaminated syringe. d

(ii) The cause of the infection is probably due to a vet using a contaminated syringe. ✗ e

e **4/9 marks awarded** a Student B refers to a 'helical primary structure' in the normal protein, which is incorrect. Helices are of course part of the secondary structure and so the mark cannot be awarded. This mistake is not carried forward and 1 mark is awarded for pleated sheets in the abnormal form. b This has been well learned, for 2 marks. c Even with the misspelling student B should have remembered CJD and so got the order of words correct (in which case the mark might have been awarded, since the spelling is phonetically close). No mark awarded. d These are all methods by which infection can be transmitted or acquired. Only 1 mark gained. e There is no reasoning here. Another animal would have to have been infected for the syringe to be contaminated. No mark awarded.

Question 5 Nucleic acids

(a) The diagram below shows part of a DNA molecule in the process of replication.

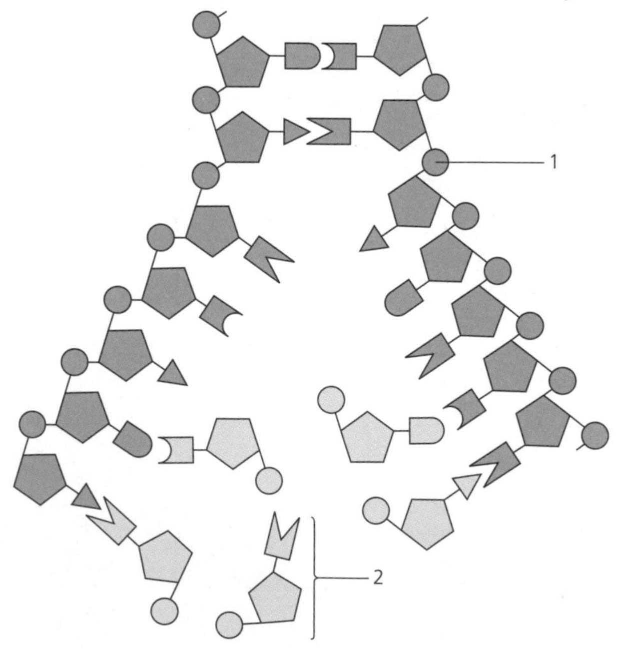

(i) Name the structures 1 and 2. (2 marks)

(ii) Name the enzymes that are responsible for

- the unzipping of the DNA molecule
- the attachment of structure 2 to the DNA strand (2 marks)

(b) Meselson and Stahl devised an experiment to determine whether DNA replication is conservative or semi-conservative. They grew bacteria for many generations in a medium containing heavy nitrogen (^{15}N) and then transferred them to a medium containing normal nitrogen (^{14}N).

Samples of bacteria were removed from the medium containing ^{15}N, and then again when they had been growing for one and two generations in the medium containing ^{14}N.

The DNA was extracted from each sample and then centrifuged. The diagram below shows some of the results.

Sample 1
Bacteria grown in medium containing ^{15}N

Sample 2
Bacteria grown for one generation in medium containing ^{14}N

Sample 3
Bacteria grown for two generations in medium containing ^{14}N

(i) Explain whether the results support the conservative or semi-conservative hypothesis of DNA replication. (2 marks)

(ii) Complete the diagram to show the position of the centrifuged DNA for Sample 3. (2 marks)

(c) In a DNA molecule 22% of the bases are adenine. Determine what percentage of the bases are guanine. Show your working. (2 marks)

(d) Describe how the bases in RNA differ from those in DNA. (1 mark)

Total: 11 marks

e Part (a) is straightforward. You should give precise names for the structures in (a) (i) and for the enzymes involved in DNA replication in (a) (ii). For part (b) you should be familiar with the Meselson and Stahl experiment, though you should take time to read all the information carefully as you will need it when answering the questions. Notice that in (b) (i) you are asked to *explain* the result with respect to both hypotheses, not just decide which is supported by the result. Part (c) involves a problem-solving exercise involving base pairing, while (d) asks for a *specific* difference between DNA and RNA.

Student A

(a) (i) 1 — sugar-phosphate backbone ✗; 2 — nucleotide ✓ **a**

(ii) DNA helicase ✓; DNA polymerase ✓ **b**

(b) (i) A single band in sample 2 is consistent with the semi-conservative hypothesis, which would result in first-generation DNA being composed of a ^{15}N strand and a ^{14}N strand. ✓ The conservative hypothesis would result in two bands, one band containing heavy ^{15}N DNA and one light, ^{14}N — this is obviously not the case. ✓ **b**

(ii) **Sample 1** **Sample 2** **Sample 3** **b**

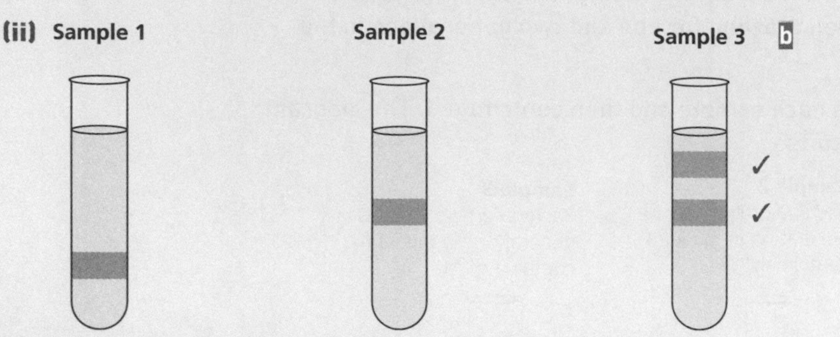

(c) 22% adenine would mean that 22% is thymine. ✓ So 56% of the DNA would contain cytosine and guanine, half of which, 28%, would be guanine. ✓ **b**

(d) While thymine is found in DNA, it is replaced by uracil in RNA. ✓ **b**

e 10/11 marks awarded **a** 1 is phosphate — the whole line of sugars and phosphates would need to be labelled for this to be identified as the 'backbone'; 2 is correct, for 1 mark. **b** All correct, for 9 marks.

Student B

(a) (i) 1 — phosphate ✓; 2 — ATP ✗ **a**

(ii) helicase ✓; polymerase ✗ **b**

(b) (i) According to the semi-conservative hypothesis the DNA would contain half of the heavy DNA and half of the light DNA so would form a single band. ✗ **c**

(ii) **Sample 1** **Sample 2** **Sample 3** **d**

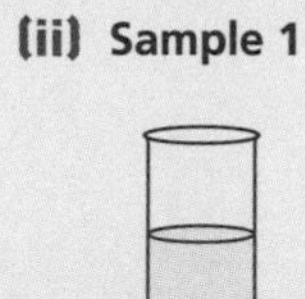
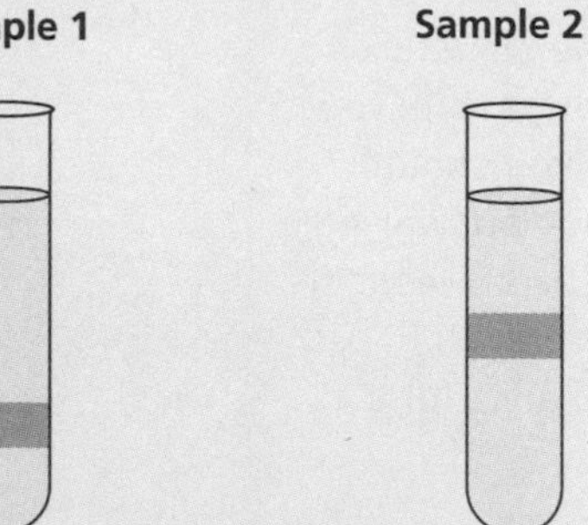
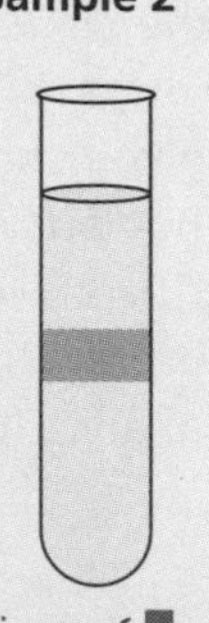
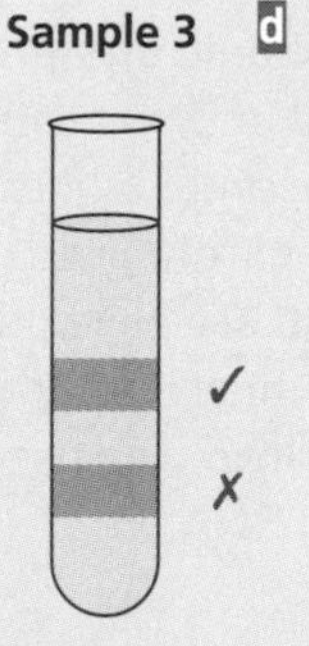

(c) There would be 28% guanine. ✓ **e**

(d) In RNA the thymine base is replaced by uracil. ✓ **f**

e 5/11 marks awarded **a** Phosphate for structure 1 is correct. Structure 2 has a base (possibly adenine), a deoxyribose but only one phosphate, so cannot be ATP. Student B earns 1 mark for naming structure 1 correctly. **b** Helicase is allowed, though DNA helicase is more precise. However, there are different types of polymerase, so student B should have specified DNA polymerase. 1 mark scored. **c** The semi-conservative hypothesis would result in a single (intermediate) band but the question asked for an explanation. The answer given is not precise. A single (intermediate) band is produced because one strand is heavy and one strand is light — just saying half heavy and half light could mean that each strand contained some of each. Also, student B has not explained why the result does not support the conservative hypothesis — there would be two bands, one heavy (with ^{15}N only) and one light (with ^{14}N only). Student B fails to score. **d** Two bands would be expected, one level with the band in sample 2 and one above this level, so only one of these is correct. The lower band is wrong — an all heavy DNA band would not be expected because the bacteria are using ^{14}N to make their new DNA. So half of the new DNA will contain all ^{14}N and half will contain equal amounts of ^{14}N and ^{15}N. 1 mark gained. **e** This is correct, but no working has been shown, as asked for in the question. Student B does not necessarily understand the base pairing rules, for example A pairing with C and T pairing with G would give the same answer. Also, student B should have noticed that the question was worth 2 marks. 1 mark awarded. **f** This is correct, for 1 mark. However, student B could have made it clear that thymine is a base found in DNA.

Enzymes

Question 6 Enzyme action

(a) The diagram below represents the arrangement of the amino acids in the enzyme to which the substrate has attached.

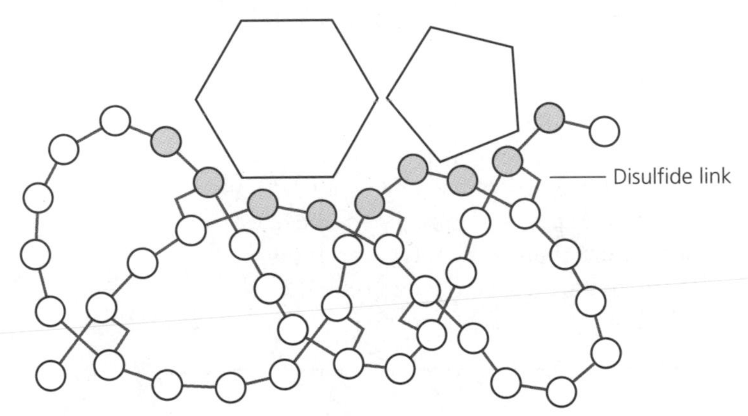

(i) Name the part of the enzyme in which the amino acids are shaded grey. (1 mark)

(ii) Use the diagram and your understanding of enzyme action to explain how an enzyme acts as a catalyst. (3 marks)

(b) **The graph below shows the amount of product formed by an enzyme-controlled reaction over time.**

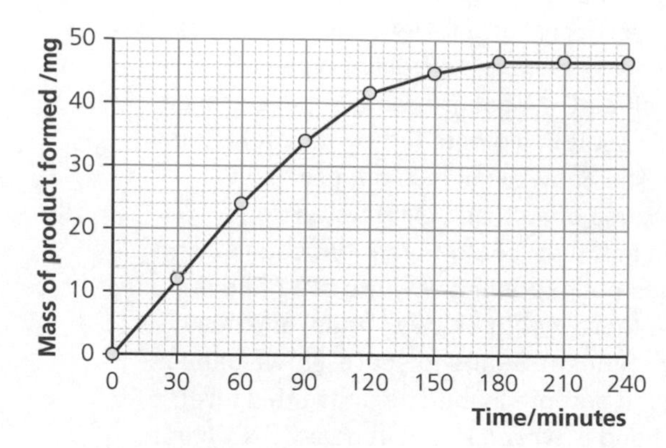

(i) **Calculate the rate of reaction, in mg min^{-1}, in the first hour.** (1 mark)

(ii) **Identify the trends evident in the graph.** (2 marks)

(iii) **Explain the trends identified.** (2 marks)

(iv) **Explain how you would control two other variables in this experiment.** (2 marks)

(c) **The chemical mercaptoethanol breaks (reduces) disulfide links. If the mercaptoethanol is then removed and the enzyme exposed to an oxidising environment, the disulfide links are reformed. Explain two ways in which the effect of mercaptoethanol on the activity of an enzyme differs from the effect of heating the enzyme to temperatures above 45°C.** (2 marks)

Total: 13 marks

e Part (a) relates to enzyme structure and function and, while it involves mostly recall, you should study the diagram carefully. In (a) (ii) you will benefit from using the diagram to explain how enzymes act as catalysts. Part (b) (i) involves a simple calculation, but make sure that you follow the instruction and answer in the unit required. In part (b) (ii) you must identify trends in a graph of enzyme activity and then, in (b) (iii), explain the trends. The main issue with a product/time graph is that students often mix this graph up with the rate of reaction/substrate concentration graph — always study the variables carefully when presented with a graph. When giving an explanation try to use the word 'because' in the first line of your answer before continuing with a reason. Aspects of experimental design are tested in (b) (iv). Notice that you must do more than just list two variables — you must say how you would control them. In part (c) you need to read the information carefully to determine the effect of the chemical specified.

Student A

(a) **(i)** active site ✓ a

(ii) The reaction takes place on the active site, which has a complementary shape to that of the substrate molecule. ✓ Enzymes effectively lower the activation energy needed for the reaction ✓ by orientating the substrates in such a way as to facilitate bonding between them or, in this case, manipulating the substrate to allow the bond to readily break. ✓ b

(b) (i) 24 mg in 60 minutes = 0.4 mg min^{-1} ✓ c

(ii) The amount of product formed increases steadily for the first 2 hours. ✓ After this, little product is formed and after 3 hours no more is produced. ✓ d

(iii) Because over the first few hours there was a large concentration of substrate so more enzyme–substrate complexes were formed and more product released. ✓ After 3 hours there was no more product formed. ✗ e

(iv) A pH buffer should be used to keep pH constant. ✓ Temperature should be controlled using a water bath, with the temperature being monitored using a thermometer. ✓ f

(c) Mercaptoethanol only affects disulfide bonds, while heating has an adverse effect on other bonds, such as hydrogen and ionic bonds, that hold the tertiary structure of the enzyme. ✓ At high temperature the enzyme is permanently denatured, while the effect of mercaptoethanol is not permanent, since disulfide bonds can re-form when the chemical is removed. ✓ f

e **12/13 marks awarded** a Correct, for 1 mark. b This excellent answer gains all 3 marks. c Correct, for 1 mark. d Both trends are described clearly, for 2 marks. e This is correct as far as it goes, for 1 mark. However, student A has forgotten to explain *why* no more product is formed — no more product is formed as all the substrate has been converted into product. f Complete answers, for full marks.

Student B

(a) (i) active site ✓ a

(ii) They speed up the rate of reaction without being altered themselves. Enzymes bind molecules called substrates with the same shape as the active site and promote the reaction that changes the substrate to products. ✗ b

(b) (i) 24 mg per hour ✗ c

(ii) From time 0 to 120 minutes the rate of reaction is high and the amount of product formed increases. ✓ After this there is no effect. ✗ d

(iii) There is a high substrate concentration initially, so the rate of reaction is high over 120 minutes. ✓ After that, the active sites are full. ✗ e

(iv) Temperature and pH should be controlled. ✗ f

(c) The enzyme would be killed by mercaptoethanol. ✗ Above 45°C bonds within the enzyme break and the enzyme becomes inactive. ✗ g

e 3/13 marks awarded **a** Correct, for 1 mark. **b** Student B defines a catalyst without explaining how it operates as a catalyst; and refers to the active site and substrate having the same shape, which is incorrect. The answer should refer to the complementary nature of the active site and substrate and explain how the activation energy is lowered. No mark awarded. **c** Student B has not looked carefully enough at the unit required and provided a rate per hour instead of min^{-1}. No mark scored. **d** The first point is correct, for 1 mark. Regarding the second point, saying that there is no effect is too vague — student B could have said that no further product is formed. **e** While the first point gains a mark, student B has failed to study the variables on the graph sufficiently well, confusing it as a rate of reaction against substrate concentration graph. The answer is that, over time, all the substrate has been converted to product. **f** These are variables that should be controlled, but the question asks *how* they might be controlled. **g** Since enzymes are not themselves alive, they cannot be killed. The effect of heating is explained, but there is no attempt to suggest how the heating effect differs from the effect of mercaptoethanol. No mark scored.

Question 7 Enzyme inhibitors

(a) (i) The diagram below represents an enzyme, its substrate and an inhibitor of the enzyme.

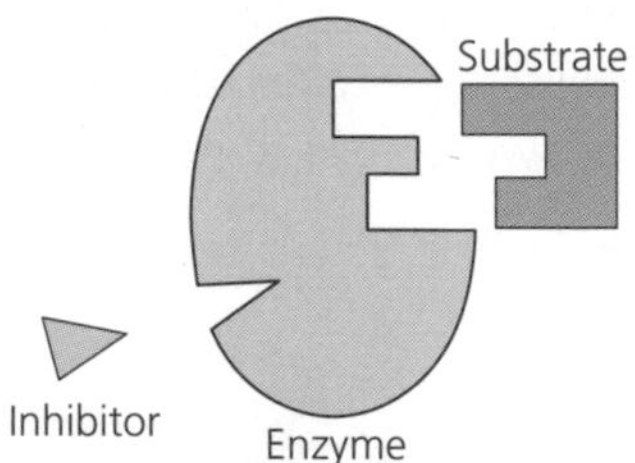

Explain whether the inhibitor shown is a competitive or non-competitive inhibitor. (1 mark)

(ii) The graph below shows the effect of increasing substrate concentration on the rate of an enzyme-controlled reaction when inhibitors X and Y are present.

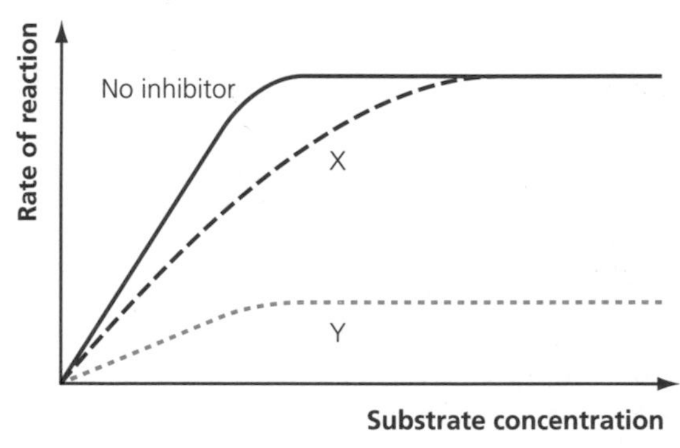

Describe and explain the differing effects of X and Y. (4 marks)

(b) **Prostaglandins are produced in the body when there is damage to tissue, and results in inflammation and pain. The enzyme that is involved in the pathway leading to the production of prostaglandins is also involved in the pathway leading to the production of thromboxane. This is a substance that promotes blood clotting. The metabolic pathway is shown below.**

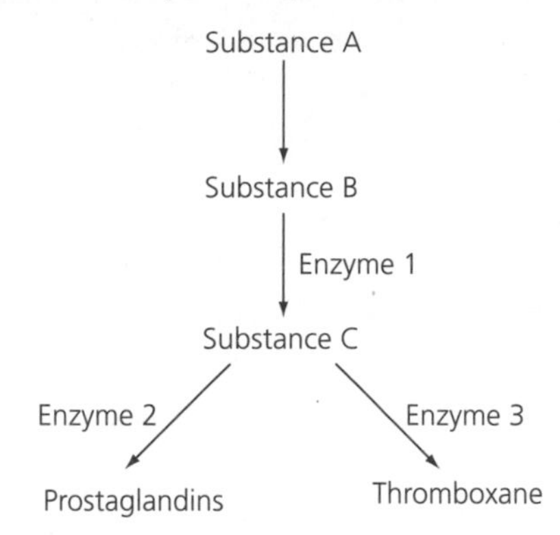

(i) **Aspirin acts as an inhibitor of one of the enzymes in the pathway. Which enzyme, 1, 2 or 3, is affected? Explain your answer.** (2 marks)

(ii) **Suggest why aspirin does not affect the other two enzymes.** (2 marks)

(iii) **When a person has a heart attack, a blood clot in a coronary artery stops blood from reaching part of the heart muscle. Suggest why people who have had a heart attack take aspirin as a treatment.** (2 marks)

Total: 11 marks

e Part (a) (i) introduces the concept of enzyme inhibition and is relatively simple. Part (a) (ii) is a more challenging question as students often find this graph difficult to interpret. This is a *describe and explain* question, so even if you are struggling to fully explain what is happening you can still pick up marks by describing what the graph is showing. Part (b) presents you with information about the influence of aspirin on metabolic pathways. You will not have seen this before, so it is assessing your ability to apply your understanding. You must study the information fully and try to work out what is happening.

Student A

(a) **(i)** The inhibitor is attaching to a part of the enzyme that is not the active site. ✗ a

(ii) As substrate concentration increases the effect of X is reduced and at high substrate concentration it has negligible effect. ✓ This is because X is a competitive inhibitor and at high substrate concentration there is a greater chance of a substrate colliding with the enzyme's active site than of an inhibitor blocking the active site. ✓

The non-competitive Y attaches to the enzyme and causes the active site to be distorted. ✓ This prevents the substrate from binding there no matter what the substrate concentration is and so the rate of reaction remains low. ✓ b

(b) (i) X ✓ because it is involved in both pathways ✓. **b**

(ii) Aspirin only attaches at a particular site with specific amino acids. ✓ Enzymes 2 and 3 do not possess this sequence of amino acids, so aspirin cannot attach. ✓ **b**

(iii) Aspirin inhibits the production of thromboxane, which promotes blood clotting. ✓ Therefore blood clotting is less likely to take place and so coronary arteries will not be blocked. ✓ **b**

e **10/11 marks awarded** **a** Student A has failed to identify the inhibitor as competitive or non-competitive and appears to have rushed into an explanation. This emphasises the importance of reading questions carefully and answering fully. No mark awarded. **b** All correct, for a total of 10 marks.

Student B

(a) (i) This is a non-competitive inhibitor since it does not attach to the active site, but binds with the enzyme at another site. ✓ **a**

(ii) Inhibitor X reduces the rate of reaction at low substrate concentration but as substrate concentration increases its effect is reduced. ✓ Inhibitor Y causes a low rate of reaction at all substrate concentrations. ✓ **b**

(b) (i) Enzyme 2 ✗, because it is involved in the production of both prostaglandins and thromboxane ✓. **c**

(ii) Aspirin cannot attach to any site on the other two enzymes. ✓ **d**

(iii) Aspirin is taken to reduce the risk of a heart attack because it prevents inflammation and fever. ✗ **e**

e **5/11 marks awarded** **a** A good answer, for 1 mark. **b** Student B has not attempted any *explanation* of how either inhibitor influences the rate of reaction at different substrate concentrations and, indeed, may not fully understand the topic. Nevertheless, he/she has made accurate observations of the graph and has provided suitable descriptions of the effect of X and Y, for 2 marks. **c** Enzyme 2 is only involved in the synthesis of prostaglandins and this may be a mistake in writing down the answer. However, the points are marked independently and so the second mark is awarded. **d** This is not a detailed enough answer though it is worthy of 1 mark. **e** Student B has not studied the information in the question stem carefully enough and has ignored the information on thromboxane. No mark awarded.

Question 8 Immobilised enzymes

(a) **The graph below shows the rate of reaction catalysed by an enzyme at different pHs, when it is free in solution (●) and when it is immobilised (○).**

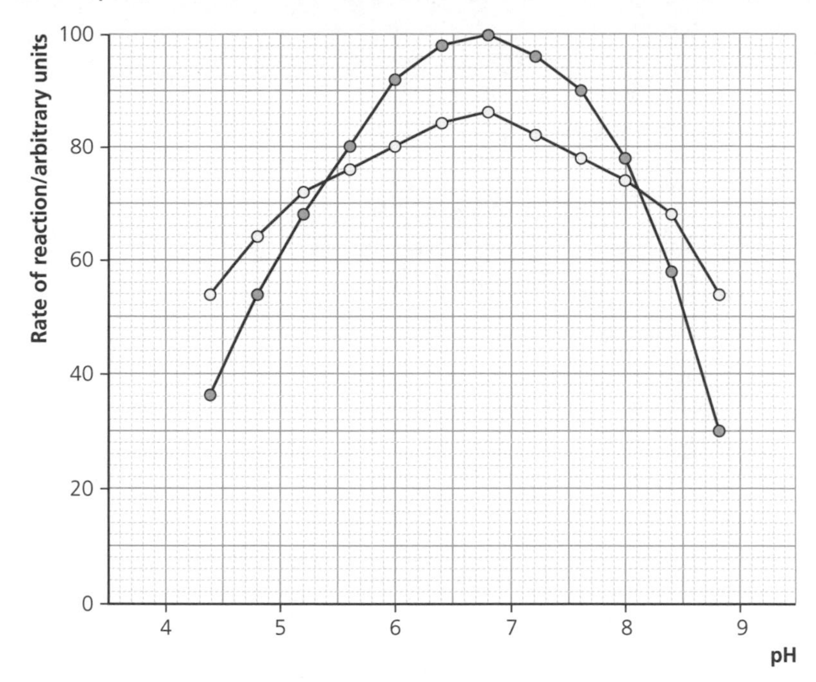

(i) **Determine the optimal pH for the enzyme.** (1 mark)

(ii) **Explain the effect of varying pH on the activity of the free enzyme.** (2 marks)

(iii) **Describe two effects of immobilisation on the rate of reaction.** (2 marks)

(iv) **Suggest reasons for the effects that you have described in (iii).** (2 marks)

(b) **Give an example of a biosensor containing an immobilised enzyme and outline how the biosensor functions.** (2 marks)

Total: 9 marks

ⓔ This question requires you to apply your understanding of immobilised enzymes. It is important in questions involving graphs to study them carefully. Part (a) (i) requires you to read accurately, from the graph, the optimal pH and explain the influence of pH in (a) (ii). The activity of the enzyme, free in solution and immobilised, is then compared: part (a) (iii) asks you to describe differences — to turn the differences shown in the curves into words — while (a) (iv) asks for explanations — reasons for the differences. Part (b) is relatively simple if you have learned the topic. Notice that you are asked to *outline*, so you are not expected to go into detail.

Student A

(a) **(i)** pH 6.6 ✗ **a**

(ii) At the optimum pH the enzyme's active site is at its most complementary for the binding of the substrate. ✓ At extremes of pH the ionic bonds are affected, altering binding at the active site. ✓ **b**

(iii) The immobilised enzyme has a lower rate of reaction at the optimum pH. ✓
The immobilised enzyme is more active at the extremes of pH. ✓ c

(iv) Immobilised enzymes are physically bound within a substance (e.g. alginate) so the substrate cannot move as freely to the enzyme. ✓ Also, if adsorbed the enzyme might wash away.
Immobilised enzymes are more stable because they are bound onto a support material. ✓ d

(b) A glucose-specific test strip to test for the presence of glucose. ✓ A glucose-specific dipstick, such as Clinistix, uses glucose oxidase and peroxidise along with a colour dye, which changes colour if glucose is present. ✓ e

8/9 marks awarded a Incorrect. Always try to make time to check through numerical work. This is obviously a slip — nevertheless, no mark can be awarded. b An excellent answer, scoring both marks. c Both observations are correct, for 2 marks. d Full explanations, for 2 marks. e A suitable example of a biosensor and an outline of how it operates are given, earning 2 marks.

Student B

(a) (i) pH 6.8 ✓ a

(ii) At the optimum pH the active site of the enzyme has the same shape as the substrate. ✗ At extremes of pH the shape of the active site is altered and the enzyme is denatured. ✗ b

(iii) The enzyme has a lower maximum rate of reaction when immobilised. ✓
The immobilised enzyme is active over a greater range of pHs. ✗ c

(iv) When immobilised, some of the active sites might not be available as they may be bound to the support material. ✓
When immobilised, the enzymes are more thermostable. ✗ d

(b) The pregnancy test, which uses an enzyme-linked antibody. ✓ e

4/9 marks awarded a Value correctly read off the *x*-axis. b The first point is incorrect because the active site does not have the same shape as the substrate — it is complementary in shape to the substrate. For the second point, it is not sufficient to say that the enzyme is denatured. A reference to the influence on ionic bonds is expected at this level. c The difference in the maximum rate of reaction is correct.

However, the second point is not correct — both free and immobilised enzymes are active over the range shown. The immobilised enzyme exhibits a *greater rate of reaction at extremes of pH*. d The first answer is correct and scores 1 mark. In the second point student B has ignored that the variable here is pH not temperature. e The first answer is sufficient for 1 mark but, for the second mark, there should have been some further information, such as that the presence of hCG is being tested for.

Cells and viruses

Question 9 The eukaryotic cell

The photograph below is an electron micrograph of part of a eukaryotic cell.

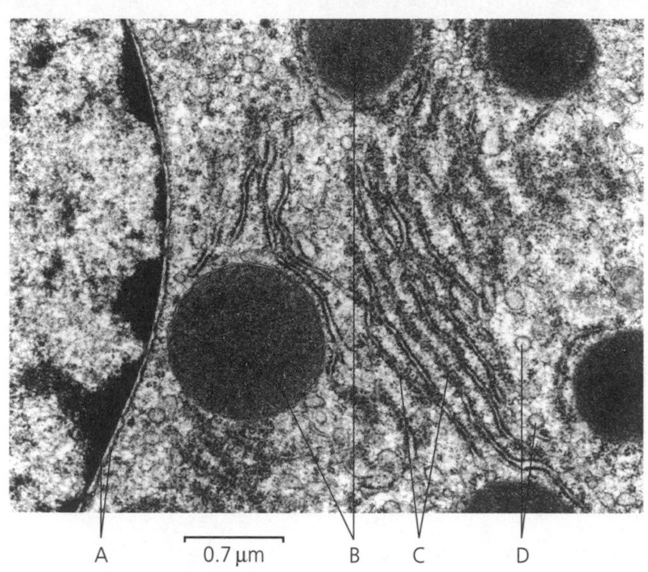

(a) **Identify the features labelled A to D.** (4 marks)

(b) **The photograph has a scale bar indicating 0.7 µm. Use this to calculate the magnification of this electron micrograph. Show your working.** (3 marks)

(c) **Explain why the internal membranes of a cell cannot be seen using a light microscope.** (1 mark)

Total: 8 marks

e Quite distinct skills are required in this question. Electron micrographs are always different from each other and, in part (a), you are required to identify four features within the specific EM provided. In part (b) you have to calculate the magnification of the EM, a skill that you should have previously practised. Part (c) is testing your understanding of microscopy — be careful about your wording when answering.

Student A

(a) **A** nuclear membranes ✓
B mitochondria ✓
C rough endoplasmic reticulum ✓
D ribosomes ✗ **a**

(b) The scale bar is 14 mm long ✓, which is 14 000 µm ✓, so magnification is 14 000/0.7 = 20 000 times ✓ **b**

(c) An optical microscope has low resolution and cannot distinguish between the membranes within the cell. ✓ **c**

e 7/8 marks awarded a The answer to D is incorrect. Ribosomes are solid structures and also smaller, as is apparent on the rough endoplasmic reticulum where the ribosomes are attached to sheets of membrane. The other answers are correct, so this scores 3 marks. Notice that the answer to A states the plural 'membranes'. b All stages in the calculation (measurement of scale bar, unit conversion and determination of magnification) are correct and clearly shown. Showing each stage is important because if a slip is made at any stage, marks can still be awarded for the correct procedure. This gains all 3 marks. c Correct, for 1 mark.

Student B

(a) **A** nuclear membrane ✗ a

B lysosomes ✗ b

C rough ER ✓ c

D vesicles ✓ d

(b) length of scale bar = 1.4 cm long ✓ × 1000 = 1400 µm ✗

magnification = 1400 × 0.7 = ×980 ✗ e

(c) The light microscope does not provide the required magnification. ✗ f

e 3/8 marks awarded a This is the nuclear envelope or double membrane, so saying 'membrane' is not sufficiently accurate. b The answer to B is incorrect. The internal cristae are not so obvious in this micrograph (though visible in places), but the features are clearly covered by an envelope, so B are mitochondria. c Notice that ER is an accepted abbreviation for endoplasmic reticulum. d Correct. e Student B has measured the length of the scale bar correctly. However, measuring in centimetres has led to an incorrect conversion to micrometres. Furthermore, if the measurement had been divided by 0.7 (the true length of the scale bar) instead of multiplying by 0.7, then another mark would have been scored. So, while the answer is wrong, 1 mark is scored because the 'working' clearly shows one correct operation. f Student B has made a common error and referred to magnification instead of resolution, which is the ability to discern fine detail. No mark awarded.

Question 10 The prokaryotic cell and viruses

(a) The diagram below shows the structure of a typical prokaryotic cell (a bacterium).

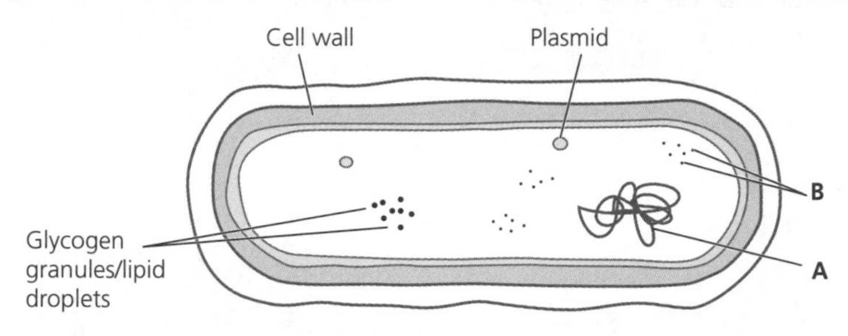

(i) Identify the structures labelled A and B. (2 marks)

(ii) Suggest one function of each of the following structures found in the bacterial cell:
- **glycogen granules**
- **cell wall** (2 marks)

(b) Mitochondria are absent from prokaryotic cells but are present in eukaryotic cells. Describe the structure and function of mitochondria. (2 marks)

(c) Viruses occur in a variety of forms, including bacteriophages (phages) and the human immunodeficiency virus (HIV).

(i) List the molecules of which all viruses are composed. (2 marks)

(ii) Explain why viruses are not regarded as living cells. (1 mark)

Total: 9 marks

e Part (a) (i) requires you to 'identify' two structures of a bacterial cell and part (ii) to 'suggest' functions for two features. You should use your understanding of animal and plant cells to help you. You are asked to 'recall' the structure and function of mitochondria in part (b). In part (c) you need to 'list' the chemical components of viruses in (i) and, while the answer to (ii) may be obvious, you need to provide sufficient detail.

Student A

(a) (i) A — DNA ✓; B — vesicles ✗ a

(ii) Glycogen granules — store glucose for respiration. ✓

Cell wall — a structural role in preventing osmotic bursting of the cell. ✓ b

(b) Mitochondria are surrounded by two membranes, with the inner membrane folded to form cristae. ✓ They contain a matrix with small ribosomes and a circular DNA molecule. In aerobic respiration much of the cell's ATP is produced. ✓ b

(c) (i) protein ✓; nucleic acid ✓ b

(ii) Viruses have no cellular structure or metabolic activity, and can only replicate using the metabolism of a host cell. ✓ b

e 8/9 marks awarded a A is correct, for 1 mark. B can only really be ribosomes — they cannot be vesicles because a prokaryotic cell has no membrane-bound organelles. b Excellent answers, for full marks.

Student B

(a) (i) A — chromosome ✗; B — ribosomes ✓ a

(ii) Glycogen granules — to store food. ✗
Cell wall — to provide support. ✓ b

(b) Mitochondria have an envelope of two membranes, the inner being folded to form cristae. ✓ They are used to produce energy. ✗ c

(c) (i) protein ✓; DNA ✗ d

(ii) They need a host to do anything. ✗ e

e 4/9 marks awarded a The first answer is not correct because bacteria possess 'naked' chromosomes consisting of DNA but lacking the associated histones. The second is correct. b Glycogen is a store of glucose (energy) — suggesting that it is a store of food is too vague. The cell wall does provide support and a mark is awarded. However, student A provides a better answer. c The structure is sufficiently well described, for 1 mark. However, it is not precise enough to say that mitochondria produce energy. They produce ATP through aerobic respiration. d All viruses contain protein, but not all contain DNA. Some, such as HIV, contain RNA. Student B earns 1 mark for protein. e This could be describing any parasite, so fails to score.

Membrane structure and function

Question 11 Water potential of plant tissue

An experiment was carried out to determine the water potential of potato tuber tissue. The procedure was as follows:

1 **Obtain tissue samples of approximately equal size.**

2 **Cut each of the samples into slices.**

3 **Surface-dry slices using filter paper.**

4 **Weigh tissue slices for each sample.**

5 **Add one sample to each of a series of sucrose solutions of known solute potential.**

6 **Leave for 24 hours, then surface-dry slices again and reweigh.**

7 **The change in mass is expressed as a percentage of the initial mass.**

(a) (i) Describe how tissue samples of approximately equal size can be obtained.

(ii) Explain why the tissue sample is cut into slices.

(iii) Explain why the slices of tissue are surface-dried.

(iv) Explain why the change in mass is expressed as a percentage. (4 marks)

(b) The results of the experiment are shown in the graph below.

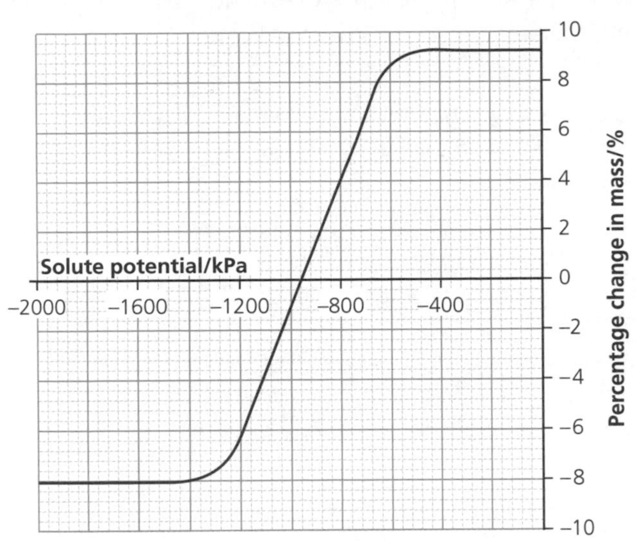

(i) Explain the change in mass of potato tissue when it was immersed in a sucrose solution of solute potential −400 kPa. (2 marks)

(ii) Determine the water potential of the potato tissue. Explain the reasoning for this determination. (2 marks)

(c) In an evaluation of the experiment the importance of a 'standardised drying technique' was emphasised. Describe what is meant by a standardised drying technique, and explain its importance to the accuracy of the result. (2 marks)

Total: 10 marks

e This question involves an investigation that you might have done in class — determining the water potential of potato tuber tissue. In part (a), and further in part (c), you are required to evaluate, and do so precisely, aspects of the procedure. In answering part (b) you need to take time to study the graph of results — read the axes carefully and do not be put off by the negative scale. When you understand the graph, you are ready to explain the osmotic changes in (i) and then explain the determination of water potential in (ii).

Student A

(a) (i) Use a cork borer of the same diameter and cut the cylinders to equal lengths. ✓

(ii) To provide a greater surface area over which osmosis can take place. ✓

(iii) To remove excess surface water that is not within the potato tissue. ✓

(iv) To allow changes in mass to be directly compared, since the tissue samples would not be the same mass to start with. ✓ **a**

(b) (i) With a solution of −400 kPa the potato gains mass and so water must have entered it ✓, which means that the potato tissue had a lower water potential ✓. **b**

(ii) 960 kPa. ✗ At this point there is no percentage change in mass, which means that the water potential in the tissue and the sucrose solution is equal ✓, so there is no net movement of water. **c**

(c) The same pressure has to be applied to the filter paper when drying before both initial and final weighings. ✓ If not sufficiently dried before the final weighing, then the weight loss would be less than it should have been. ✓ **d**

e 9/10 marks awarded **a** These are excellent, well-phrased answers for the full 4 marks. **b** Both points are well explained, for 2 marks. **c** Student A has omitted the negative sign, though has provided a correct rationale, for 1 mark. **d** Excellent answers, for 2 marks.

Student B

(a) (i) By using a cork borer. ✗

(ii) To increase the surface area for exchange. ✓

(iii) To remove any water from the surface that may affect the results. ✓

(iv) To make it easier to see the difference. ✗ **a**

(b) (i) Water moves from an area of high water potential to an area of low water potential ✓ and so moves out ✗ of the potato. **b**

(ii) −880 kPa. ✗ The pressure potential is zero, so the water potential is equal to the solute potential. ✗ **c**

(c) The same pressure is applied to the filter paper when drying the potato discs, ✓ so that the results are reliable. ✗ **d**

e 4/10 marks awarded **a** The use of a cork borer alone is not sufficient because the cylinders have to be cut to the same length. The second and third answers, though correct, might have been more fully explained (see student A's responses). The fourth answer is incorrect. 2 marks are scored. **b** Movement of water from high to low water potential is correct, for 1 mark. However, water does *not* move out of the potato tissue, since at −400 kPa there in an increase in the mass of the potato. **c** Student B has not accurately read the *x*-axis scale and needs to check his/her numerical work. The reasoning given is not that for determining the water potential, but for determining the solute potential of, for example, epidermal tissue. Student B has confused the two experiments and fails to score. **d** The description of the standardised drying technique is correct, for 1 mark. However, its use has nothing to do with reliability, which is often confused with accuracy.

Question 12 Active uptake of ions

(a) Carriers in the cell-surface membrane are responsible for the active uptake of potassium ions and other solutes. The diagram below represents the mechanism.

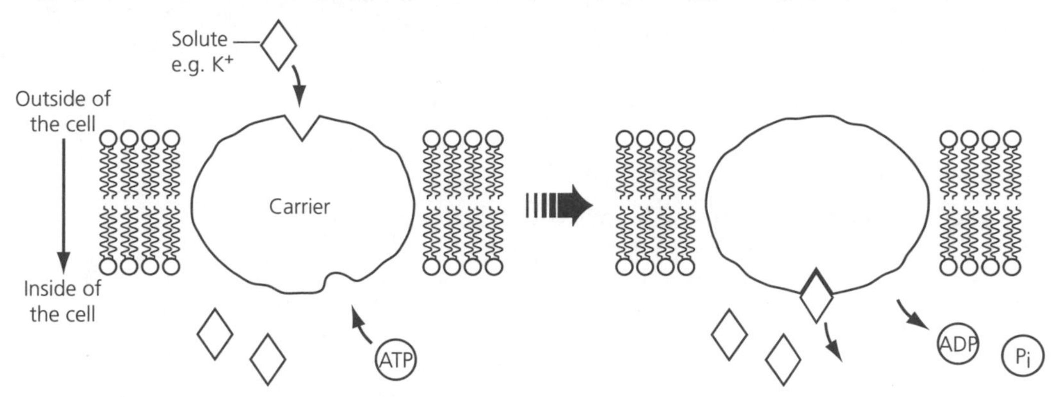

(i) What type of molecule is the carrier? (1 mark)

(ii) For each type of solute that is actively absorbed there is a specific membrane carrier. Explain this specificity. (2 marks)

(iii) Explain how water passes through the membrane. (2 marks)

(b) An experiment was carried out to investigate the uptake of potassium ions. Plant tissue was added to potassium chloride solution of different concentrations and at two different temperatures. The results are shown in the table below.

Initial concentration of potassium chloride solution/mM	Uptake of potassium ions at 5°C/arbitrary units h^{-1}	Uptake of potassium ions at 20°C/arbitrary units h^{-1}
0	0	0
5	14	30
10	18	38
20	22	48
40	23	50

(i) Plot the above data, using the most appropriate graphical technique. (4 marks)

(ii) Explain why the rate of potassium ion uptake is greater at 20°C than at 5°C. (2 marks)

(iii) Suggest an explanation why, at 20°C, increasing the concentration of potassium ion:
- **from 0 mM to 20 mM greatly increases the rate of uptake**
- **from 20 mM to 40 mM has negligible effect on the rate of uptake** (2 marks)

(iv) Rubidium ions and potassium ions have similar chemical properties. If rubidium is added to the immersing solution, the rate of potassium ion uptake is reduced. Suggest a reason for this. (1 mark)

Total: 14 marks

e Parts (a) (i) and (ii) test your understanding of membrane carriers. Be careful in (a) (iii), there are 2 marks available because there are two routes for the passage of water through a membrane. Part (b) (i) requires the construction of a graph. Generally in AS papers you are required to undertake a skills activity — where you have to *do* something rather than recall facts or work something out. Marks are awarded for:

- writing an explanatory caption
- choosing the correct type of graph, with the independent variable on the *x*-axis
- labelling the axes correctly and choosing an appropriate scale for each
- plotting the data correctly, appropriately joined and with a key

Part (b) (ii) requires you to 'explain', while (iii) and (iv) ask you to 'suggest' — i.e. provide a reasonable explanation using the information supplied.

Student A

(a) (i) protein ✓ **a**

(ii) Each carrier protein has a receptor site ✓ that has a complementary shape ✓ for the attachment of a specific solute. **b**

(iii) Water molecules are sufficiently small to diffuse across the phospholipid bilayers ✓ and through water channel proteins called aquaporins ✓. **c**

(b) (i) **d**

The uptake of potassium ions in different concentrations of KCl solution at two different temperatures, 4°C and 18°C, by plant tissue

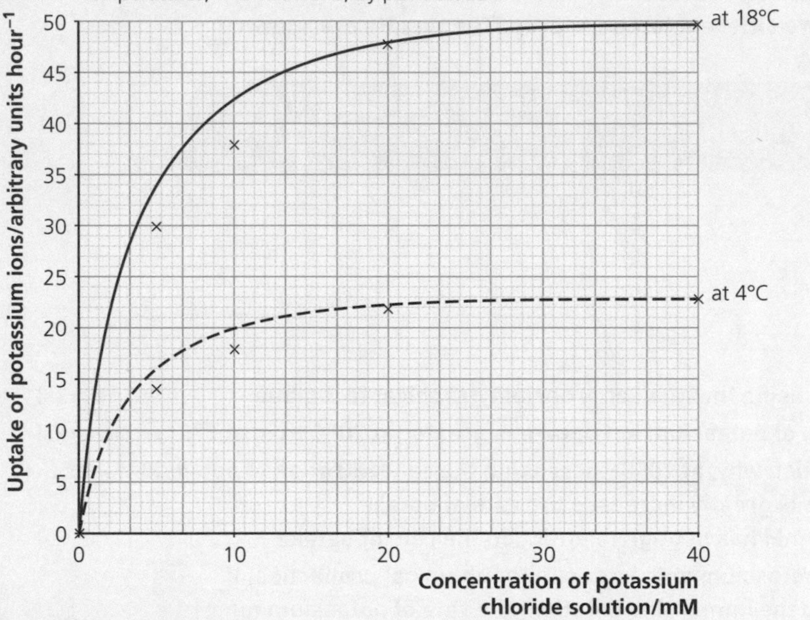

Uptake of potassium ions/arbitrary units $hour^{-1}$

50, 45, 40, 35, 30, 25, 20, 15, 10, 5, 0

at 18°C

at 4°C

0, 10, 20, 30, 40

Concentration of potassium chloride solution/mM

(ii) At higher temperature the rate of respiration is greater ✓, so more ATP is available for the action of the carriers in active transport ✓. **f**

(iii) From 0 mM to 20 mM: there are more ions available for transport. ✗

From 20 to 40 mM: the carrier proteins are functioning at their maximum rate. ✓ **g**

(iv) Rubidium ions inhibit respiration. ✗ **h**

 **11/14 marks awarded** a This is correct, for 1 mark. b This scores both marks. c Correct, for 2 marks. d An appropriate caption noting K^+ uptake, potassium chloride concentration and temperature is included ✓; a line graph is drawn with the concentration of KCl solution as the independent variable ✓; both axes have labels with units of measurement and are appropriately scaled ✓; the points are accurately plotted but joined by curved lines, which are not best-fit ✗. Student A scores 3 marks. e Student A has correctly identified that active transport requires ATP, and that the rate of respiration is dependent on temperature. This scores both marks. f The answer to the first part should have specified the frequency of attachment of the ions to the carrier proteins. The answer to the second part is correct for 1 mark. g This would reduce the rate of K^+ ion uptake, but the answer does not use the information provided — that rubidium ions have similar properties to potassium ions.

Student B

(a) (i) protein ✓ a

(ii) Fat-soluble molecules go through the phospholipid bilayer, and water-soluble molecules go through the hydrophilic channels. ✗ b

(iii) Through the hydrophilic channels. ✗ c

(b) (i)

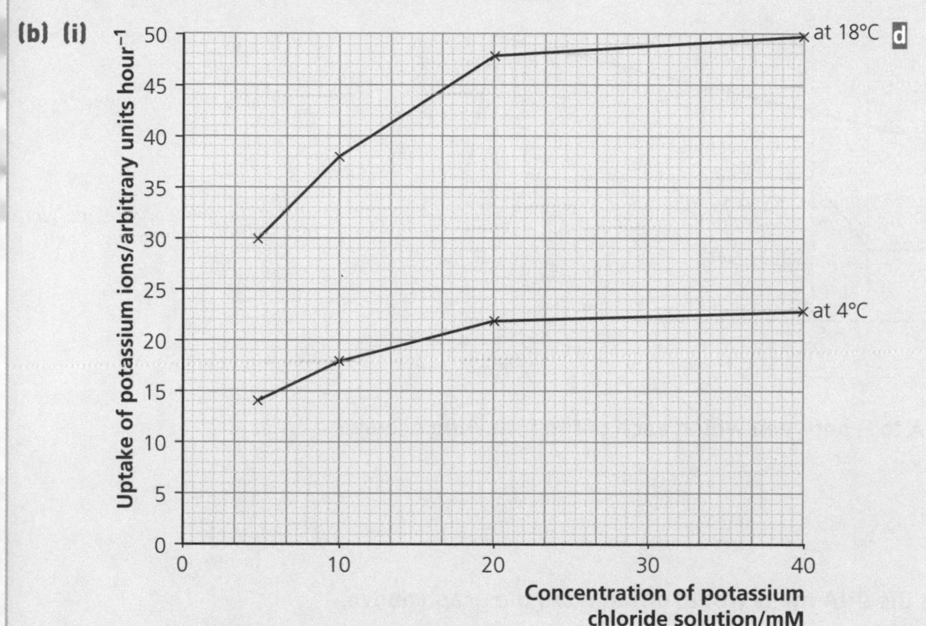

(ii) As temperature increases, the rate of ion uptake increases because the ions have more kinetic energy. ✗ e

(iii) From 0 mM to 20 mM: more potassium ions are available for attachment to the carriers. ✓

From 20 to 40 mM: there are a limited number of carrier proteins to transport the potassium ions. ✓ f

(iv) They inhibit the carrier proteins by taking the place of K^+ ions. ✓ g

e 6/14 marks awarded **a** This is correct, for 1 mark. **b** Student B has not read the question carefully enough and fails to score. The question requires an explanation for the specificity of membrane carriers. **c** This is not sufficiently precise — water molecules pass through specific water channel proteins *and* are small enough to pass through the phospholipid bilayer. No marks scored. **d** There is no caption to explain the contents of the graph ✗; a line graph is chosen and the concentration of KCl solution is the independent variable ✓; both axes have labels with units of measurement and are appropriately scaled ✓; the 0,0 points are not plotted and so the initial part of the graph is missing ✗ (though points are appropriately joined with short, straight lines and the lines for 4°C and 18°C are identified); 2 marks scored. **e** In (b) (ii) it was established that ion uptake does not occur by diffusion. The answer should relate to the availability of ATP for the operation of the membrane carriers (see student A's response). No marks awarded. **f** The answers to both parts of the graph are correct, for 2 marks. **g** This is correct, for 1 mark.

The cell cycle, mitosis and meiosis

Question 13 The cell cycle and cancer

(a) The graph below shows the change in cell mass and the change in DNA mass per nucleus in a mitotic cell cycle.

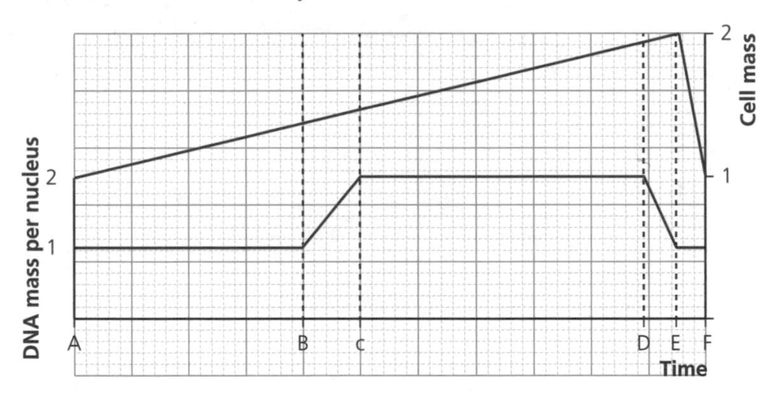

(i) Give the two letters, A to F, between which each of the following occurs:
- **G_1 phase**
- **S phase**
- **telophase**
- **cytokinesis** (4 marks)

(ii) State how changes in the DNA mass would differ, from the graph above, during a cell cycle involving meiosis. Explain your answer. (2 marks)

(b) Three checkpoints control the cell cycle. Describe what happens at each checkpoint:

(i) G_1

(ii) G_2

(iii) metaphase (3 marks)

(c) The p53 protein has an important regulatory function in the G_1 checkpoint. A mutation of the gene that codes for p53 production is associated with many cancers. Suggest why. (2 marks)

(d) Mikanolide is a drug that inhibits the enzyme DNA polymerase. Explain why mikanolide may be effective in the treatment of cancer. (2 marks)

(e) Many drugs that are used to treat cancer have side effects, such as hair loss and anaemia. Suggest why. (2 marks)

Total: 15 marks

e The graph in (a) is unlikely to be familiar and so is challenging. It is vital that you think carefully about the cell cycle and relate your understanding to what the graph is showing — notice that the DNA mass is per nucleus, not per cell. Part (b) tests your knowledge of the cell cycle checkpoints and should be straightforward. Part (c) is a 'suggest' question — you are not expected to know about p53. You need to *apply* your understanding of what happens when a checkpoint fails — the answer requires two relevant points about how cancer is formed. In part (d) you need to work out the effect of the drug on the progress of the cell cycle. The mark tariff shows that you need to make two clear points for a complete answer. Part (e) is another 'suggest' question — you are asked to work out appropriate answers.

Student A

(a) (i) G_1 — A–B ✓; S — B–C ✓; telophase — D–E ✓; cytokinesis — E–F ✓ **a**

(ii) The pattern would be the same up until point F when there would be the further division of meiosis II, ✓ producing haploid cells with a DNA mass of 0.5. ✓ **a**

(b) (i) At G_1 there is a check for DNA damage. If there is damage the cell enters a resting stage called G_0 while the DNA is repaired. ✓

(ii) At G2 there is a check that DNA has been replicated without damage. ✓

(iii) During metaphase there is a check that chromosomes are attached to the spindle fibres via centromeres and have aligned across the equator. ✓ **a**

(c) The gene mutation will result in malformed or no p53 protein being produced. In the absence of normal p53 protein cells divide uncontrollably and form tumours. ✓ The tumour becomes cancerous when cells spread to other organs forming further tumours. ✓ **a**

(d) DNA polymerase is required for DNA replication. ✓ **b**

(e) The drugs that attack rapidly growing cancer cells also attack rapidly growing cells in the body. ✓ Presumably, rapidly growing cells are involved in the production of hair and blood cells. ✓ **c**

e **14/15 marks awarded** **a** All correct, for 11 marks. **b** This is only worthy of 1 mark. If DNA replication fails then the S phase cannot be completed and so cell division is arrested. **c** Student A has worked out an excellent answer to gain 2 marks.

Student B

(a) (i) G_1 — A–B ✓; S — B–C ✓; telophase — D–F ✗; cytokinesis — E–F ✓ **a**

(ii) DNA mass would fall again as the cell would divide further so that eventually homologous chromosomes and chromatids are separated. ✓ **b**

(b) (i) G_1 — DNA checked for damage and any damage repaired. ✓

(ii) G_2 — again DNA is checked for damage. ✗

(iii) Metaphase — check that spindle has formed properly. ✗ **c**

(c) Lack of p53 may mean that a check fails and that cell division continues uncontrollably. ✓ **d**

(d) Drugs are poisons that damage normal cells. ✗ **e**

(e) Drugs cannot tell the difference between cancer cells and healthy cells that are dividing. ✓ **f**

e 7/15 marks awarded **a** Student B seems to think that telophase overlaps with cytokinesis, which is wrong. 3 marks are awarded. **b** This point is well made. However, the examiners want a more precise answer — that the eventual mass of DNA would be 0.5. Only 1 mark awarded. **c** The G_1 checkpoint is the only one that is described in any detail. Easy marks are being thrown away here through not learning straightforward facts. 1 mark scored. **d** This answer is not sufficiently detailed for both marks, for example there is no distinction between uncontrolled growth and cancer. Student B should have noticed that two marks are available so two points are required. **e** This is too vague and barely relevant. There is no attempt to work out what would happen if DNA polymerase was inhibited. Student B fails to score. **f** This is fine for 1 mark but is not full enough — student B has failed to suggest the involvement of hair follicle cells or of cells dividing to produce new blood cells.

Question 14 Mitosis and meiosis

(a) The drawing below shows stages of mitosis.

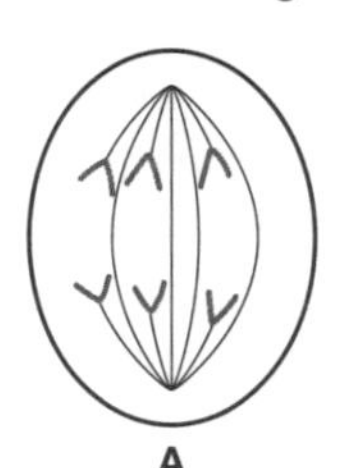
A

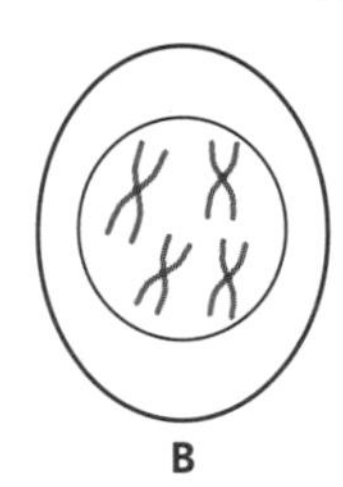
B

C

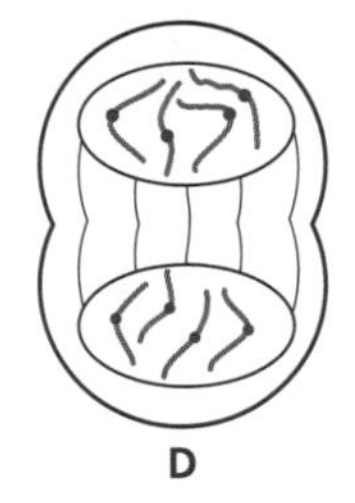
D

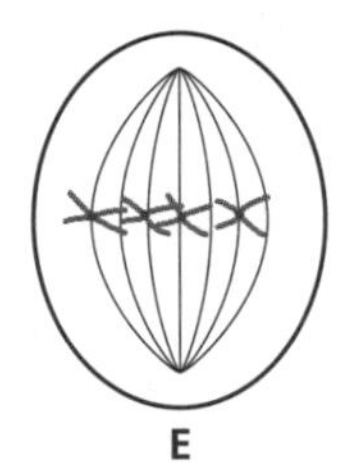
E

(i) Arrange the letters, A to E, to give the correct sequence of stages during mitosis, starting with stage C. (1 mark)

(ii) Outline what happens during stage A. (3 marks)

(b) Describe how mitotic cell division in plant cells differs from this type of division in animal cells. (4 marks)

(c) Genetic variation is generated during meiosis through the processes of crossing over (recombination) and independent assortment. Describe the events that lead up to each process and state the stage of meiosis at which each occurs.

(i) crossing over (recombination) (2 marks)

(ii) independent assortment (2 marks)

Total: 12 marks

e Part (a) tests your understanding of mitosis, while in (b) you must provide a detailed list of differences in the cell division (not just mitosis) of plant and animal cells. Questions on how meiosis generates genetic variation, as in (c), are always difficult, generally because you have to be particularly careful with your wording.

Student A

(a) (i) CBEAD ✓ a

(ii) Microtubules attach to the centromeres ✓, disassemble, tubulin protein is released at the centromere end ✓, pulling the chromatids apart ✓. b

(b) Cytokinesis in animal cells involves cleavage, whereby elements of the cytoskeleton pull the cell membrane inwards. ✓ Plant cells divide through vesicles coalescing along the equatorial plate. ✓ Also, plant cells do not possess centrioles. ✓ c

(c) (i) Crossing over occurs during prophase I ✓ as a result of chiasmata forming and an exchange of parts of chromatids from homologous chromosomes, so that different combinations of alleles are produced ✓.

(ii) Independent assortment occurs during metaphase I ✓ as the homologous chromosome pairs are arranged randomly on the equator of the spindle so that during anaphase I any one chromosome will be segregated along with any members of other homologous pairs. ✓ d

e **11/12 marks awarded** a Correct, for 1 mark. b A concise answer, scoring 3 marks. c This is good with regard to cytokinesis, but a full answer requires more detail about centrioles, for example that plant cells still produce a spindle without their presence. 3 marks scored. d Full and well-worded answers, for 4 marks.

Student B

(a) (i) CEADB ✗ a

(ii) Chromosomes are pulled apart ✗ as the spindle fibres shorten. ✓ b

(b) Animal cells possess centrioles, which they use to produce a spindle during mitosis. ✓ Plants cell lack centrioles and another mechanism is used to organise tubulin in the production of a spindle. ✓ c

(c) (i) Crossing over results in new combinations of alleles as portions of non-sister chromatids are swapped. ✓ It takes place during prophase. ✗

(ii) Independent assortment takes place during metaphase. ✗ The homologous pairs line up any way on the equator so that when pulled apart a huge variety of chromosomes combinations may be formed. ✓ d

e 5/12 marks awarded **a** Student B has not recognised B as late prophase. Stage D, which is telophase, would be followed by interphase by which time chromosomes have dissembled. No mark awarded. **b** Firstly it is not chromosomes that are pulled apart but chromatids. Student B has not been careful enough about the use of these terms. Also, there is no mention of the chromatids being pulled to the poles via their centromeres. 1 mark scored. **c** Student B has a thorough knowledge of spindle formation. Unfortunately, cytokinesis has been ignored and so the answer is incomplete. 2 marks scored. **d** The descriptions of the two processes are adequate for 2 marks. However, student B has lost easy marks by not specifying that the events take place during meiosis I, i.e. prophase I and metaphase I respectively,

Tissues and organs

Question 15 The ileum

The photomicrograph below is of a transverse section through part of the wall of the small intestine (ileum).

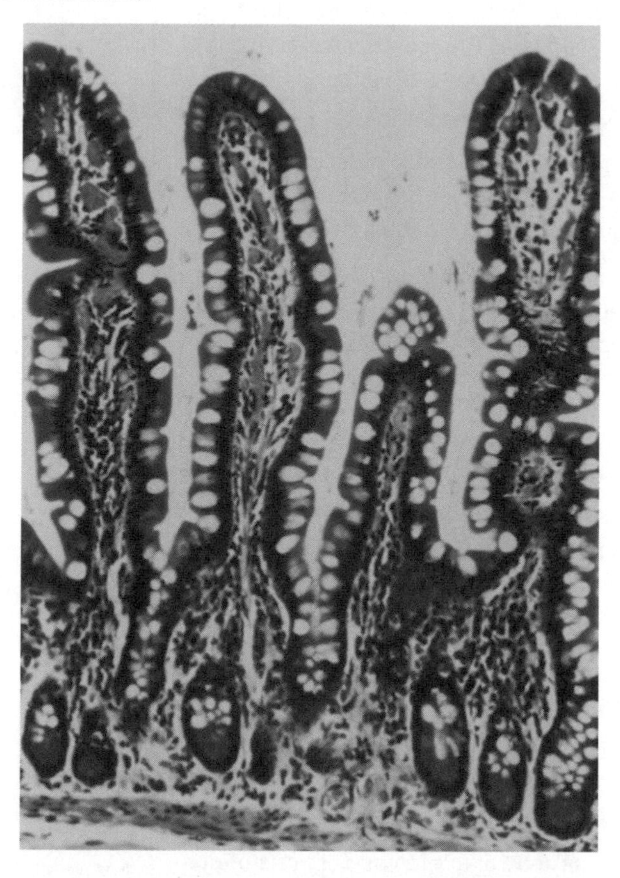

Draw a block diagram to show the tissue layers in the ileum, as shown in the photograph. Label the drawing to identify at least five structures. (9 marks)

Total: 9 marks

ⓔ This question involves *drawing* — the skills activity in this paper. You also have to identify five structures. Notice that the outer layers of the ileum are not included in the photomicrograph. There are 4 marks for drawing skills and 5 marks for identification. Drawing skills marks are awarded for drawing the obvious tissue layers, ensuring that your drawing is that of the photo supplied, that the proportionality of the drawing is accurate and that the lines are clear and not sketchy.

Student A

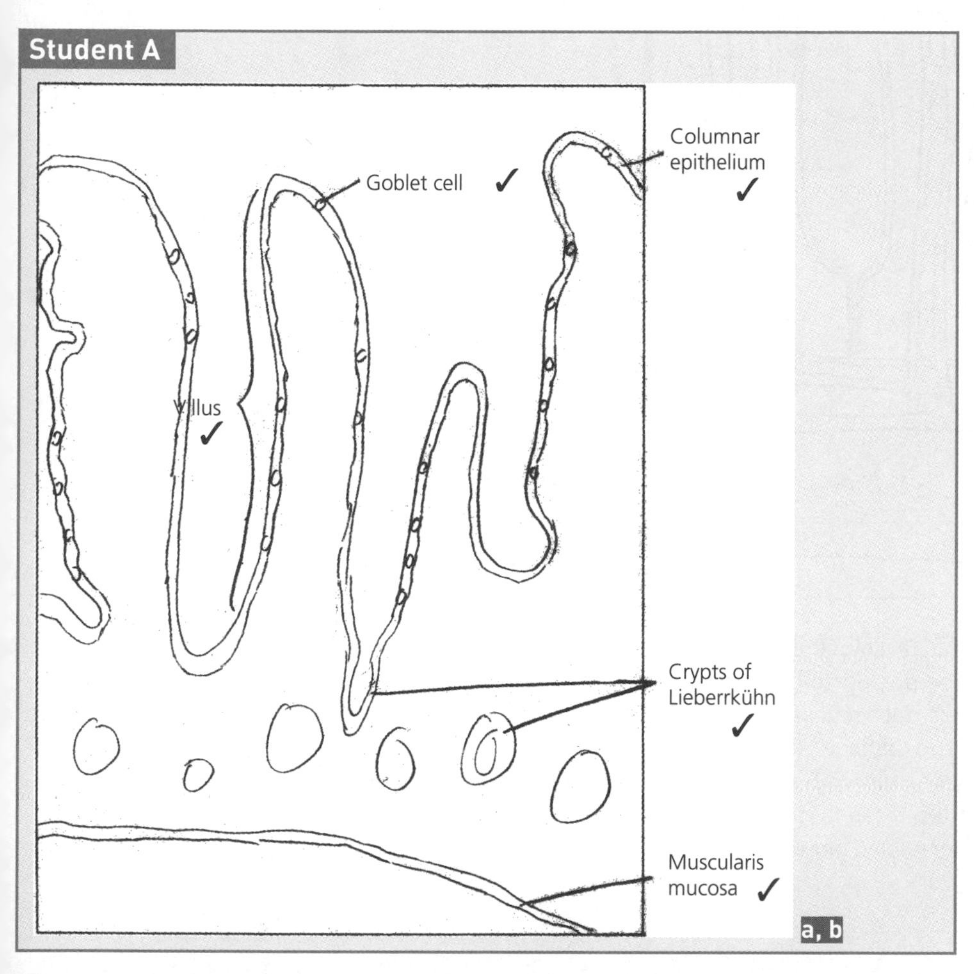

ⓔ **8/9 marks awarded** a This is a block drawing showing tissue layers that illustrate all the obvious features. ✓ It is a fair attempt to draw the photograph ✓ and the proportionality of features is sufficiently accurate. ✓ However, the lines tend to be sketchy in places, and circular structures, such as those for the goblet cells, are incomplete. ✗ Student A earns 3 marks out of 4 for drawing skills. b Five labels are correctly identified, for 5 marks.

Student B

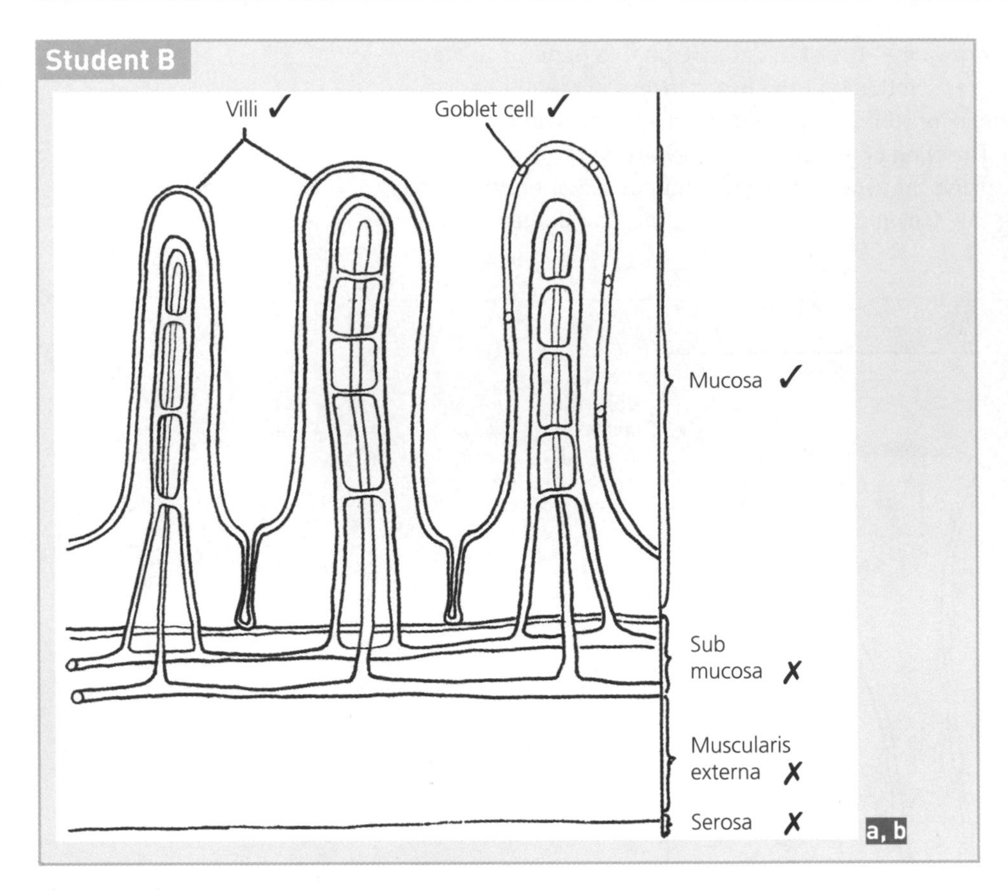

e **5/9 marks awarded** **a** This is a block diagram showing the tissue layers of those tissues obvious in the photograph. ✓ However, it is a well-learned textbook diagram, which does not accurately represent the photograph ✗ and lacks the proportionality of the features shown. ✗ The lines drawn are smooth and continuous, not sketchy. ✓ Student B earns 2 marks out of 4 for drawing skills. **b** Student B has labelled three features correctly, for 3 marks. The layers submucosa, muscularis externa and serosa are not included in the photograph and so cannot be awarded marks.

Question 16 The leaf

The diagram below shows a transverse section across part of a leaf.

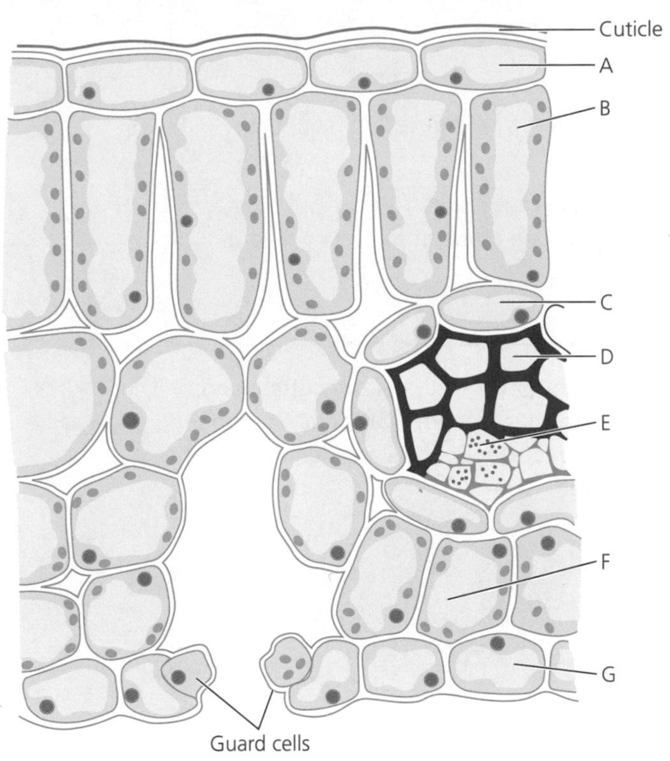

The letters A–G identify different tissues in the leaf.

(a) **Identify the tissue in which the following processes take place:**

- **(i)** **maximum light absorption** (1 mark)
- **(ii)** **diffusion of gases** (1 mark)
- **(iii)** **transport of sugars** (1 mark)

(b) **(i)** **State the role of the cuticle.** (1 mark)

- **(ii)** **The guard cells have a dual role in the leaf. Explain this role.** (2 marks)

(c) **Plant cells are cemented together by a middle lamella. State one ion that the plant must take up to produce the middle lamella.** (1 mark)

Total: 7 marks

e This is a straightforward question. Part (a) requires you to recognise tissues in the leaf and recollect their function. You are asked to recall the role of two leaf structures in (b) — notice that in (ii) you must provide two separate points. In (c) you need to remember the roles of ions in the synthesis of a plant cell structure.

Student A

(a) (i) B ✓

(ii) F ✓

(iii) E ✓ a

(b) (i) The waxy cuticle reduces the loss of water by evaporation from the surface of the leaf. ✓

(ii) Guard cells open during the day to allow gases to diffuse in and out of the leaf (CO2 in for photosynthesis) ✓ and close at night to prevent excessive transpirational loss of water. ✓ b

(c) nitrate ion for protein synthesis ✗ c

e 6/7 marks awarded a All correct, for 3 marks. b The roles of the cuticle and stomata have been well described. A total of 3 marks gained. c Student A seems to have forgotten about calcium or magnesium pectate. Pectins are derivatives of polysaccharides, not proteins. No mark awarded.

Student B

(a) (i) B ✓

(ii) F ✓

(iii) D ✗ a

(b) (i) The cuticle waterproofs the leaf. ✗ b

(ii) Guard cells regulate the rate of transpiration by opening and closing the stomata. ✗ c

(c) calcium ion ✓ d

e 3/7 marks awarded a The first two answers are correct. Student B has confused xylem (D) with phloem (E) — the phloem always lies underneath the xylem in the leaf. b This is not precise. If the cuticle simply waterproofed the leaf then there would be no need for a thicker cuticle on the upper surface. It is more accurate to say that the cuticle reduces evaporation from the surface. No mark. c This answer is simply too vague to be worthy of a mark. d Correct, for 1 mark. Magnesium ion would also have been allowed.

Section B Essay questions

Question 17 Polysaccharides

Give an account of the structure and function of polysaccharides. (15 marks)

In this question you will be assessed on your written communication skills.

Total: 15 marks

ⓔ You should spend some time organising an essay *plan* and ensure that this relates to both structure and function of each of the different types of polysaccharides. The essays are marked according to the number of worthy (called indicative) points wihin the content and placed in one of three bands; the final mark is then based on the quality of written communication. Quality of written communication is assessed on your ability to sequence your ideas, emphasise the links between structure and function and on the use of appropriate specialist vocabulary.

Student A

Starch is the energy store in plants ✓ and is made up of amylose and amylopectin. ✓ Both are polymers of α-glucose. ✓ They are helical molecules and so compact ✓, while amylopectin is branched due to 1,6-glycosidic bonds ✓.

Glycogen is the energy store in animals ✓, particularly in the liver and muscle tissue ✓. It is also a polymer of α-glucose ✓ and is similar to amylopectin in that it is branched due to the presence of 1,6-glycosidic bonds, though branching occurs more frequently ✓. The many terminal ends mean that it can be rapidly hydrolysed. ✓

Cellulose is the structural polysaccharide found in the plant cell wall. ✓ It is a polymer of β-glucose ✓ and so forms straight chains ✓. The cellulose molecules hydrogen bond together ✓ forming microfibrils ✓ of high tensile strength ✓. **a**

Quality of written communication **b**

ⓔ **15/15 marks awarded** **a** 16 marking points are identified. **b** The detail and balance within the essay is good, and the many valid points place the indicative content in the top band. The student expresses ideas clearly and fluently through well-sequenced sentences, while the use of biological vocabulary is good throughout, so that a total of 15 marks is awarded.

Student B

Starch is usually found in plants. It is an energy store in plants ✓ and its glucose isomers are composed from α-glucose ✓. It is a branched structure and has no hydrogen bonds present. Glycosidic bonds are present. **a**

Glycogen is found in the liver and muscles ✓ of animals. It is the energy store ✓ and its glucose isomers are composed of α-glucose and β-glucose. It is a branched structure and also has glycosidic bonds present. **b**

Cellulose is found in the cell walls in plants. ✓ It has many functions, which include a food source, a structural component of the cell wall and for tensile strength. It is composed of β-glucose ✓ and glycosidic bonds are present. **c**

Quality of written communication **d**

ⓔ **7/15 marks awarded** **a** Two points are given. Other information is not sufficiently detailed — for example, starch cannot be described as branched when branching only occurs in its amylopectin component. **b** Two points are provided. It is wrong to say that glycogen is a polymer of α-glucose and β-glucose — it is a polymer of α-glucose. Detail is also lacking — branching should be explained as due to 1,6-glycosidic bonds. **c** Two points are identified. Other phrases lack detail — for example, tensile strength is not explained and is not by itself sufficient to earn a mark. **d** With 6 indicative points the essay is placed in the middle band. While there is reasonable use of biological terms, links between structure and function are not well established, so a total of 7 marks is awarded.

Question 18 The cell-surface membrane

(a) Describe the structure of the cell-surface (plasma) membrane. (8 marks)

(b) Explain how membrane structure determines how molecules pass through the membrane. (7 marks)

In this question you will be assessed on your written communication skills.

Total: 15 marks

e In part (a) you should include all aspects of the cell-surface membrane structure. A *plan* will help. Part(b) is more discriminating, and a *plan* will be even more important if you are not to miss some important aspects of membrane transport. Both parts are marked according to the number of worthy points within the content and placed in one of three bands; the final mark is then based on the quality of written communication. The quality of written communication is assessed on the clarity and coherence of your writing and on the use of appropriate specialist vocabulary.

Student A

(a) The cell surface membrane consists of a bilayer of phospholipids. ✓ The heads of phospholipids are polar and hydrophilic, so they are arranged outermost in water. ✓ The tails of the phospholipids are non-polar and hydrophobic, so they remain in contact with each other. ✓ There are proteins ✓ interspersed among the fluid phospholipids, so the structure is described as a fluid-mosaic model ✓. There are intrinsic proteins that are embedded within the bilayer and there are extrinsic proteins. ✓ Some proteins have carbohydrate attached ✓ and this acts as a cell recognition feature. Cholesterol is also present among the phospholipids and stabilises the fluidity of the membrane ✓, especially when the temperature changes. **a**

(b) Small gaps in the phospholipids allow entry of small molecules such as CO_2, O_2 and water, which is also polar, by simple diffusion. ✓ Molecules that are too large and polar must pass through protein channels. ✓ Each has a site for a specific molecule so there are many different types of protein channel for different types of molecule. ✓ Carrier proteins can also change shape — a molecule binds to the protein, it changes shape and the molecule is released inside the cell. ✓ Facilitated diffusion is passive ✓, requiring no energy input. Active transport goes against the concentration gradient ✓ and requires energy in the form of ATP. ✓ Where active transport takes place, there is a large number of mitochondria. **a**

Quality of written communication **b**

e 15/15 marks awarded **a** A total of fifteen worthy points are identified. **b** With the high number of worthy points, Student A is placed in the top band for both parts. They are well-structured accounts with the ideas expressed fluently and there is good use of specialist vocabulary. The relationship between membrane structure and how substances pass across the membrane is clearly made. A final mark of 15 is awarded.

Student B

(a) The cell membrane is composed of proteins ✓ floating in a fluid bilayer of lipid. This structure is called the fluid-mosaic model. ✓ It is made of two layers of phospholipids ✓ that have heads and tails. The tails of these two layers face each other with the heads facing in opposite directions. The tails are non-polar ✓ and the heads are polar. ✓ Proteins consist of long chains of amino acids. Some of these are polar and some aren't. Some rest on the surface of the bilayer while others go right through. **a**

Quality of written communication **b**

(b) Proteins may act as carriers. Other proteins act as enzymes. Cells have to be recognised by antibodies and hormones because of proteins on the surface. Water is able to pass between the phospholipid molecules of the bilayer because they are very small. ✓ Water-soluble substances also use this route. Glucose, which is polar, relies on carrier proteins. ✓ Facilitated diffusion takes substances against the concentration gradient. **c**

Quality of written communication **d**

e 8/15 marks awarded **a** Five appropriate points are given. In some cases, however, the marks are only just arrived at. 'Bilayer of lipid' is not sufficient to earn a mark, through a following sentence describes 'two layers of phospholipids'. Marks are awarded over two sentences for polar heads facing outmost and non-polar tails facing innermost. The comment about some of the amino acids in the proteins being polar and some being non-polar is not sufficient to earn a mark. Student B should have stated that the amino acids in contact with the lipid layer are non-polar. Further, in the last sentence it is not clear whether 'some' refer to proteins or amino acids. **b** Student B is placed in the middle bad for indicative content. Since the account is well expressed, though fails to use the full range of specialist terms to describe membrane structure, a final mark of 6 is awarded for part (a). **c** Two appropriate points are provided. Some points are simply wrong, for example lipid-soluble substances (not water-soluble) pass directly through the bilayer while facilitated diffusion moves substances down (not against) the concentration gradient. Other points are not relevant: proteins acting as enzymes or hormone receptors have nothing to do with movement across the membrane. **d** This places Student B in the lower band, and since the account strays from the topic and lacks the biological terms expected in an explanation of membrane transport, part (b) is awarded 2 marks.

Knowledge check answers

1 a Water molecules are charged and so surround ions and other charged molecules, separating them in solution.

b Molecules that lack a charged group, i.e. non-polar molecules (e.g. lipids), will not dissolve in water.

2 heat

3 $C_7H_{14}O_7$

4 Starch molecules can differ in the relative amounts of amylose and amylopectin, and in the number of glucose molecules contained in each.

5 Being more highly branched means that glucose can be more rapidly hydrolysed from glycogen. This is important since animals have relatively high rates of respiration (metabolism) and so use up glucose quickly.

6 Amylose is a polymer of α-glucose, cellulose of β-glucose; amylose is helical, cellulose forms a straight chain; β-glucoses are inverted in cellulose, α-glucoses are not inverted in amylose.

7 A triglyceride is not composed of repeating sub-units — it has three fatty acids attached to a glycerol.

8 A triglyceride consists of three fatty acids bonded to a glycerol, while a phospholipid consists of two fatty acids and a phosphate bonded to a glycerol, i.e. one fatty acid is replaced by a phosphate.

9 The primary structure is held by peptide bonds; the secondary structure by hydrogen bonds; and the tertiary by hydrogen, ionic and disulfide bonds.

10 Similarities: both have a quaternary structure (consisting of more than one polypeptide); both are found in animals.

Differences: haemoglobin consists of four polypeptides while collagen has three; two types of polypeptide occur in haemoglobin while in collagen they are identical; haemoglobin is globular (has a tertiary structure) while collagen is fibrous; haemoglobin is conjugated (containing haem groups) while collagen is not; haemoglobin has a functional role (carries oxygen) while collagen has a structural role.

11 Because prion proteins are resistant to high temperatures.

12 two

13 Phosphate at one end (the 5′ end) and a pentose sugar at the other (the 3′ end).

14 Chains running alongside one another, but in opposite directions.

15 TAGACAT

16 75% light and 25% intermediate

17 a catabolic

b anabolic

18 The activation energy is the energy barrier that has to be overcome before the reaction can happen.

19 This is because the enzyme's active site has a precise shape and distinctive chemical properties so that only a particular type of substrate molecule can bind.

20 a An increase in substrate concentration increases the rate of reaction since there is a greater chance of collision with an enzyme molecule.

b An increase in enzyme concentration does not increase the rate of reaction since the enzyme is already present in excess.

21 Denaturation breaks bonds in the tertiary structure of the enzyme. This alters the shape of the active site and the substrate is unable to bind to the enzyme.

22 Little or no maltose would be produced.

23 non-competitive (and permanent)

24 *Any three from:* the rate of reaction of the immobilised enzyme is greater between 0°C and 35°C; the optimum temperature of the immobilised enzyme covers a wider range; the immobilised enzyme begins to denature at a higher temperature; the free enzyme is more active at 40°C.

25 Fructose does not fit into the active site of glucose oxidase.

26 Because neither human cells nor viruses have a cell wall (of peptidoglycan).

27 stomach bleeding and ulcers

28 ×1500

29 To keep the pH of the cell contents constant. Any change in pH could denature the enzymes in the mixture; so the function of any organelles isolated could not be studied.

30 Cut fresh leaves in cold isotonic buffer. Filter to remove cell debris. Centrifuge supernatant further: the first pellet is likely to contain nuclei, as they are the densest organelles; the second pellet will contain chloroplasts which are the next densest (and denser than mitochondria). Pour off the supernatant and re-suspend the chloroplasts in isotonic buffer.

31 a *any three from:* cellulose cell wall; plasmodesmata; sap vacuole; chloroplasts; starch grains

b cell wall of chitin; lysosomes; glycogen granules

32 a smooth ER

b mitochondria

c centrioles

33 a Golgi body

b mitochondrion

c chloroplast

34 a protein synthesis

b secretion

c absorption

35 a The nucleolus produces the components of ribosomes.

b Proteins are made on the ribosomes, enter the rough ER and are encased in vesicles, which pinch off and fuse with the forming face of the Golgi body where they are modified and again encased in vesicles, which pinch off the mature face.

36 a nucleic acid (DNA or RNA) core and a protein coat

37 Viruses lack any metabolism (e.g. respiration); lack any organelles, which specialise in metabolic activities, or cytoplasm; possess only one type of nucleic acid; cannot themselves reproduce.

38 TATACATGAG

39 Intrinsic proteins are partially or fully buried within the phospholipid bilayer. Extrinsic proteins are superficially attached to either surface of the membrane.

40 They differ in the number and types of proteins they contain. (They all have the same phospholipid bilayer.)

41 They have the specific receptor for the attachment of insulin.

42 It should be small and lipid soluble.

43 Channel proteins form hydrophilic pores, which are often shaped to allow only one type of ion through. Carrier proteins are shaped so that a specific molecule (e.g. glucose) can fit into a complementary site at the membrane surface. When the specific molecule fits, the protein changes shape to allow the molecule through to the other side.

44 Solute dissolves and water molecules cluster around the solute molecules. This reduces the capacity for water molecules to move freely and the water potential decreases.

45 The water potential will increase.

46 Both use carrier proteins; active transport requires ATP and occurs against the concentration gradient.

47 Microvilli present a larger surface, so that more carrier proteins can be supported.

48 endocytosis

49 S phase of interphase

50 During nuclear division animal cells possess centrioles (which act as a focus for spindle fibres) while plant cells lack centrioles (so spindle fibres are parallel). In animal cells cytokinesis occurs by cleavage, while in plant cells cytokinesis occurs through cell plate formation.

51 S phase, since DNA replication (DNA synthesis) cannot take place.

52 **D**, **d**, **d** (from the left)

53 Haploid, because there is an uneven number of chromosomes, which means that they cannot form homologous pairs.

54 **a** 4

b 2

55 2 (of the 4)

56 8 (2^3)

57 Because the cell-surface membranes of the epithelial cells are partly composed of proteins (e.g. involved in facilitated diffusion and active transport).

58 Contraction of longitudinal muscle causes pendular movements, while contraction of circular muscle causes local constrictions.

59 Palisade cells lie towards the upper section of the leaf towards the source of light; they are cylindrical, so reducing the number of light-absorbing cell walls; they are packed with chloroplasts, which contain photosynthetic pigments. All of these features maximise light absorption for photosynthesis.

60 Opening during the day facilitates the uptake of CO_2 for photosynthesis, while closing at night reduces the transpirational loss of water.

Index

Note: page numbers in **bold** indicate defined terms.

G

H

I

K

L

M

N

O

P